Microcomputer Applications in
Measurement

Macmillan titles of interest to engineers

Boxer *Work Out Fluid Mechanics*
Boxer *Work Out Thermodynamics*
Drabble *Work Out Dynamics*
Hall *Polymer Materials, 2nd edition*
Hull and John *Non-Destructive Testing*
Jackson and Dhir *Civil Engineering Materials, 4th edition*
John *Work Out Engineering Materials*
Mosley and Bungey *Reinforced Concrete Design, 4th edition*
Mosley and Hulse *Reinforced Concrete Design by Computer*
Simonson *Engineering Heat Transfer, 2nd edition*
Spencer *Fundamental Structural Analysis*
Stone *Introduction to Internal Combustion Engines*
Stone *Motor Vehicle Fuel Economy*
Turner *Instrumentation for Engineers*
Uren and Price *Surveying for Engineers*

Foundations of Engineering Series

Drabble *Dynamics*
Hulse and Cain *Structural Mechanics*
Powell *Electromagnetism*
Silvester *Electric Circuits*

Microcomputer Applications in Measurement Systems

C. J. Fraser
BSc, PhD, CEng, MIMechE, MInstPet
Reader in Mechanical Engineering
Dundee Institute of Technology

J. S. Milne
BSc, CEng, FIMechE
Senior Lecturer in Mechanical Engineering
Dundee Institute of Technology

MACMILLAN

First published 1990

Published by
MACMILLAN EDUCATION LTD
Houndmills, Basingstoke, Hampshire RG21 2XS
and London
Companies and representatives
throughout the world

Typeset by TecSet Ltd, Wallington, Surrey
Printed in Hong Kong

British Library Cataloguing in Publication Data
Fraser, C. J.
  Microcomputer applications in measurement systems.
  1. Science. Experiments. Scientific experiments.
  Applications of microcomputer systems
  I. Title   II. Milne, J. S.
  502'.8'5404

  ISBN 0–333–51837–3
      0–333–51838–1 Pbk

Contents

Preface

During the previous decade the most significant improvements in instrumentation and measurement have been centred round the development of microelectronic devices. The modern engineering laboratory displays an ever increasing range of microprocessor-based products: supplementing various instrumentation and primary sensors, and providing a new dimension of intelligence and control in engineering measurement systems. Information technology in the laboratory has removed the tedium associated with the data-taking process. Modern systems incorporating acquisition, logging, analysis and presentation of data are rapidly becoming the rule, as opposed to the exception. Microcomputers are therefore becoming increasingly evident in research and general experimental environments and undoubtedly they will continue to exert a growing influence in the enhancement of the basic instrumentation systems of the future. The importance of the application of microcomputers in experimentation has grown to the extent that an introductory text covering the fundamental principles would now seem to be particularly warranted.

To the uninitiated, a full understanding of the microcomputer-based experimental test facility would apparently demand a detailed knowledge of instrumentation, micro-electronics, high and low level computer programming, computer architecture, the experimental method and control engineering. With such an extensive range of sub-topics to be mastered, the task of familiarisation might seem daunting to all but the most determined. This however is not the case, since although a basic knowledge is certainly required in all of the above areas, all-round in-depth expertise is not. Against this backdrop, the scope of this text is to provide a foundation for the application of the new technology in the experimental environment. The text therefore covers all of the fundamental concepts within each of the sub-topics to a sufficient depth. This should enable the readers confidently to develop their own custom-built computerised data-acquisition and control systems and subsequently to use these systems to their fullest advantage. Detailed expositions of the finer points are omitted, although the level which is covered will provide readers with a sound foundation upon which to develop their confidence in any of the specialist associated areas. An adequate number of references to the more specialist books are included as an aid to this development.

The book is intended primarily for students following a B Eng or a Higher National Diploma course in Engineering, but should prove useful to other science students who have a major element of experimentation incorporated in their courses. Research students about to develop an automated, or partially automated, experimental test facility may find the text of particular relevance. As an introductory text to a very broad based subject, the book should also prove useful to students following the new technology courses which are being introduced in a number of secondary school syllabuses.

Numerous full worked examples are included to lead the reader onto the more complex situations and gradually to instill confidence in the use of a microcomputer for the acquisition, reduction, analysis, storage and finally the presentation of experimental data. In addition, digital control strategies are included to give a balanced coverage of the utilisation of the microcomputer in the modern experimental test facility.

It is assumed that students will be studying concurrently, or have studied, a course in computer programming in some or other dialect of BASIC. Many of the examples and illustrative material are particular to the BBC microcomputer, with the 6502 microprocessor, or to the IBM-PC and compatibles with the 8088, or 8086, microprocessor. In this manner we hope to encompass as wide a range of microcomputer hardware, relevant to the needs of industry and educational establishments alike.

Wherever possible, the semi-conductor devices and various other integrated circuits are referred to in the text by their generic type numbers. For the added benefit of UK readers, the equivalent RS Components Ltd stock reference numbers are also quoted.

Most of the programs presented, with the exception of the very short ones, are actual machine listings of working code. In this way we hope to have minimised any errors in transcription of the code.

Acknowledgements

In the preparation of this book we have been fortunate to have had the encouragement and assistance of many of our colleagues in the department of mechanical engineering. Our students, past and present, have contributed in no uncertain terms to the contents herein through their commendable efforts in project investigations and research activities. To them we owe a special vote of thanks.

We are also very fortunate to have had the very able services of messrs Ken Rusk and Bob Greig who have often transformed our scribbled and whimsical notions into quality hardware.

The support of Macmillan Education is gratefully acknowledged and particular thanks are also due to the reviewers, for their astute guidance in the preparation of the manuscript.

Lastly we would thank our wives, Ann and Isabel, for their unstinting faith that we would get it right eventually.

Dundee, 1989   C. J. Fraser
                                                                    J. S. Milne

Notation

b	breadth (mm)
d	depth (mm)
e.m.f.	electro-motive force (volts)
f	frequency (Hz), also friction factor in pipe flow in chapter 2
f_c	cut-off frequency (Hz)
f_s	sampling frequency (Hz)
h	heat transfer coefficient (kW/m^2 K), also manometer reading (mm)
i	current (amps)
k	thermal conductivity (kW/m K), also roughness height (mm)
k/D	relative roughness parameter in pipe flow (chapter 2)
$\dot{m}$	mass flow rate (kg/s)
n	integer number
p	pressure (kN/m^2 or bar)
r	resistance (ohms)
s	Laplacian operator ($\partial/\partial t$)
t	time (s), also temperature in degrees Centigrade
v	voltage (volts), also dependent variable in chapter 3
x	linear displacement (mm)
y	vertical displacement (mm)
A	cross-sectional area (m^2)
B	binary number
C	velocity (m/s), also used to denote a capacitor
C_d	discharge coefficient, drag coefficient
C_p	specific heat capacity (kJ/kg K)
D	diameter (mm)
E	error, efficiency, modulus of elasticity (N/m^2)
F	force (N)

G	gauge factor, Gray number, also used to denote a strain gauge
H	head (m)
I	current (amps)
K	calibration constant, gain
L	length (m)
M	arithmetic mean, mass (kg)
N	rotational speed (rev/min)
Nu	Nusselt number
O	observation, reading
P	dependent variable
Q	flow rate (litres/s)
R	resistance (ohms), also damping factor in chapter 8
R_0	resistance at zero degrees Centigrade
R_i	internal resistance (ohms)
Re	Reynolds number
S	spring stiffness (N/m)
T	torque (N m), also temperature in degrees Kelvin or Centigrade as specified in the text
T'	percentage turbulence
T_d	derivative time constant
T_i	integral time constant
U	mean velocity (m/s)
V	voltage (volts)
V_i	input voltage (volts)
V_o	output voltage (volts)
W	load (N)
X	numerical input
Y	numerical output
ΔR	change in resistance (ohms)
Δp	differential pressure (kN/m^2)
Δt	sampling interval (s)
Ω	ohms
β	resistivity (ohms/cm)
ϵ	strain
η	efficiency
μ	viscosity (kg/m s)
ν	Poisson's ratio
ξ	damping ratio
ρ	density (kg/m^3)
σ	standard deviation
τ	time constant (s)
ϕ	phase angle
ω	circular frequency (rad/s), also range of a variable in chapter 3

ω_c cut-off frequency (rad/s)
ω_n natural frequency (rad/s)
ω_d damped frequency (rad/s)

Subscripts
o   referred to standard conditions, also used to denote output
i   input
ref   reference quantity
x   referred to a linear displacement
a   axial
t   transverse
T   total

Prefixes
k   kilo (10^3)
m   milli (10^{-3})
M   mega (10^6)
μ micro (10^{-6})
n   nanno (10^{-9})
p   pico (10^{-12})

Chapter 1
The Engineering Laboratory

1.1 THE NEED FOR EXPERIMENTATION

The technological advances of the twentieth century continue at an accelerating pace, yet we are continually confronted with physical systems for which the mathematics are either inadequate or analytically insoluble. Turbulence provides one notable example where seven basic unknowns, comprising three component mean velocities, their fluctuating components, and the pressure are governed by only five basic equations. The experimental method, consisting of repeated trial, error, modification and refinement therefore plays a fundamental role. Until such time that the mathematical sciences can fully account for the physical behaviour of systems, experimentation will continue to play an important part in the founding of the hitherto unknown physical laws and the practical empirical relationships.

1.2 OVERVIEW OF THE EXPERIMENTAL METHOD

In essence an experiment is carried out in order to gain knowledge on the behaviour of a particular system and how it might respond to external influences. The first problem facing the experimenter is how to devise a test, or series of tests from which the required information might be derived. In an engineering context this would normally involve the construction of a piece of test equipment, varying in complexity with the scale of the problem being investigated. In most instances the test facility would represent a facsimile or a scaled model of the physical system. The next task (usually done in parallel) is to decide on the number, range and level of accuracy of the experimental variables to be measured. This is often based on experience, or at least on some prior background knowledge of the nature of the problem under study.

Supposing that the test programme is aimed at determining the response P of a physical system which is thought to be dependent on two independent variables, Q and R, that is

$$P = f(Q, R)$$

Holding say Q constant, the response of P to ten different levels of R could be measured. With a new fixed value for Q the response of P to the ten different R levels could again be measured. If ten representative values of Q are to be included in the study, a total of one hundred individual experiments must be carried out. If there is a third variable, S, in the equation, then one thousand separate tests will be needed to give the complete response, P, to ten different levels of each of the independent variables Q, R and S. It is apparent that even if the tests are relatively straightforward they will still require a considerable investment in time. The number and range of experimental variables should therefore be given very early and careful consideration.

Of similar importance is the degree of accuracy required. When Gallileo carried out his famous inclined plane experiments in the first years of the seventeenth century, he used as his timing device a system where water was collected in a measuring vessel from an elevated reservoir. Although clocks were available at that time, Gallileo considered them insufficiently precise for his exacting requirements. At the other end of the scale however, over-accuracy can be unnecessarily expensive and should be avoided wherever possible. The essential thing is to have full knowledge of the level of accuracy achieved in each measurement. Only with such knowledge can error accumulation in derived quantities be properly assessed and quantitatively presented, so that action may be taken, if necessary, to improve the situation, see chapter 3.

Armed with an extensive data bank, including uncertainty estimates, the experimenter's next task is to analyse the data, looking for trends and patterns which will allow the data to be encapsulated, hopefully in one all encompassing equation, or at best into a family of similar curves. Having achieved this happy state, the final stage in the process is the formal presentation of the findings to a satisfactory and professional level. This may involve written or oral presentation or just as likely both of these, and a structured approach is as essential here as at any other step in the sequence.

From this overview of the experimental method, the reader could be forgiven for deriving the opinion that there is nothing really new in experimentation: the same basic principles have been practised since Gallileo's time and it is proper that they should continue to be. While this observation is certainly correct, it is the means by which the experimentation is carried out that has changed so radically. In the last decade a whole new subject, loosely titled 'Microprocessor-based data-acquisition', has evolved and is progressively becoming an essential element of instrumentation, measurement and experimentation.

1.3 IMPACT OF THE NEW TECHNOLOGY IN THE ENGINEERING LABORATORY

The first emergence of computer-based data acquisition systems can be traced back to the early 1970s where a popular, but expensive, system was the Digital Equipment Corporation PDP-11. This machine was more akin to the so-called 'mini-computer' than to the modern day microcomputer. However, in addition to a generally high initial purchase cost, this particular generation of machines required considerable maintenance and they have now largely been superseded. By the end of the 1970s the introduction of the Commodore 'PET' heralded the era of the low-cost microcomputer system and the later arrival of the 'PUPI', a 12 bit analogue input/output (I/O) and digital I/O for the 'PET' user port, signalled the beginnings of inexpensive microcomputer-based data acquisition systems. The 1980s have seen tremendous development in microcomputer technology with ever increasing user memory available on most machines and ever decreasing capital costs for complete systems. Thus in the 1990s, computer technology can hardly be ignored and this is nowhere more valid than in the engineering laboratory.

One major advantage to be gained with a microprocessor-based data-acquisition system is an increase in experimental productivity. Enhanced productivity, however, is by no means the only advantage to be gained through microcomputer experimentation. The microcomputer is the implement of information technology and as such affords easy mass storage of data on magnetic disc for equally easy retrieval, transmission, re-analysis or simple record keeping. A $5\frac{1}{4}$ inch 'floppy disc' typically can hold as much as 0.4 mega-bytes of information. As every alpha-numeric character occupies one byte of storage location, then it is possible to store 0.4 million characters on the floppy disc. One page of A4 text might contain something like 250 lines of typing. With typically 80 characters per line, including the spaces, there are approximately 20,000 characters on a single typed page of A4. It is evident therefore that about 20 pages of A4 text could be stored comfortably on a floppy disc. Thus, ease of data handling, coupled with convenience and an intrinsic compactness forms the second major advantage associated with microcomputer based experimentation.

The third major advantage relates to the instrumentation itself. In the pre-microcomputer era, test facilities bristled with complementary instrumentation. Digital voltmeters for measuring mean levels of signal, r.m.s. meters for fluctuating levels, oscilloscopes for visual representation, and so on, increasing in complexity and cost with the depth of the investigation. With the arrival of the new technology however, the whole philosophy of measurement has changed. The modern day equivalent consists of some kind of transducer, which will output a voltage which is in some way related, not necessarily linearly, to the physical variable being measured. Using an analogue to digital (A/D) converter the analogue voltage is processed and represented digitally. Using fast data-acquisition techniques, this digital representation can then be 'captured' for subsequent analysis and display.

These latter functions in fact make much of the previously required instrumentation totally redundant. For example, if the mean level of the signal is required, then by simply summing all the samples taken and dividing by the total number of samples, the average value is obtained. In such a manner one expensive digital voltmeter is replaced by two or three lines of computer coding. Using the powerful graphics facilities available on most machines, the signal can be plotted on a monitor screen, amplified or attenuated via the software, re-plotted against an extended or contracted time-base and finally 'dumped' to a printer for a permanent record. These are the functions usually available on a good quality digital storage oscilloscope and all can be reproduced by relatively simple programming on a microcomputer. The level of analysis in fact is boundless, with such features as digital filtering and Fast Fourier Transform analysis being quite commonplace. It is entirely up to the users to work out their own requirements and explore the possibilities. The essential point is that many functions which required expensive instrumentation in the past can now be reasonably accommodated by software as opposed to hardware development. As a precautionary note however, it must be stated that there are still some limitations on the microcomputer-based data-acquisition system. These include the acquisition rate, the memory size available and the resolution achievable. Nonetheless, these limitations constitute areas of very active research and development and will undoubtedly continue to improve in the future.

1.4 TOWARDS THE AUTOMATED LABORATORY

The impact of the new technology in the engineering laboratory is manifested in increased productivity, improved data handling and storage techniques and the development of intelligent instrumentation. All of these come under the general umbrella of data-acquisition systems. The logical progression, and to some extent already with us, is the fully automated laboratory.

A common feature of many current data-acquisition systems and their associated microcomputer is the use of the digital I/O port to send on/off control signals to various devices. For example, to switch on a stepper motor to move a sensing device to a new measuring station, or to send a 'start recording' prompt to a data recording device. This kind of feature is the first step towards full automation. The industrial counterpart already exists in the form of system monitoring and control. The power and process industries exhibit many examples of these automated systems, from the unmanned gas production platforms in the southern sector of the North Sea to the computerised monitoring and control units found in the electricity generating complexes and in the paper making industries. Industry, in fact, is replete with such complex systems and it is only their generally high cost which has so far prevented their proliferation in all but the well funded engineeering laboratories.

1.5 CONCLUDING REMARKS

The previous decade has seen an increase in applications of the new technology in experimental investigations. In consequence, a completely new curriculum subject has developed which demands detailed attention in its own right. The technologists of the 1990s, and beyond, will need to be well versed in this technology if they are to respond to the rapid developments which are taking place and which will continue to evolve. In technical parlance, the adjective 'unmanned' is taking on a significant prominence, whether referring to deep space probes or to the projected development plans for the exploitation of the world's oil resources. It is clear therefore that system monitoring and data acquisition, combined with computer based control, will be fundamental in this development.

Chapter 2
Measurement of Primary Data

2.1 ANALOGUE AND DIGITAL TRANSDUCERS

The primary data in any experiment consists of the raw basic measurements before any manipulation or data reduction processes have been carried out. In acquiring the primary data, transducers of one sort or another will have been used. The transducer may be considered as a device which converts a measured physical variable, or measurand, into some convenient form of signal. For a microcomputer-based data-acquisition system the transducer signal must ultimately be in the form of an electrical voltage. A Bourdon-tube pressure gauge, for example, is a common enough instrument for measuring pressure, but it cannot be linked to a microcomputer unless the graduated scale indicator can be related in some manner to an applied voltage. While this could actually be done, the availability and range of relatively low-cost pressure transducers would not warrant the practice. The requirement that the output is in voltage form is a consequence of the nature of the analogue interface, see section 6.3, which performs the translation of the analogue voltage level into a digitally encoded number.

The most commonly measured parameters in an experimental context might include temperature, flow rate, pressure, speed, force, torque, power, displacement, angle, strain, humidity, light intensity, acidity, sound or simply time. The list is by no means exhaustive, but all of the given variables can, with the appropriate transducer, be relayed in terms of a proportional voltage. It is beyond the scope of this book to detail elaborate descriptions of the operating principles and characteristics of each of the many types of transducer which are currently available. As an alternative therefore, the emphasis is restricted to particular applications involving the common transducers and the reader is referred to the more specialist texts for an in-depth exposition of transducer characteristics.

Conventional transducers have an analogue voltage output which may be monitored using traditional instrumentation including for example an oscilloscope or a digital voltmeter. However, if the data is to be down-loaded to a microcomputer for subsequent analysis, it must first of all be conditioned and then digitised prior to transmission. An obvious advantage to the end user would be a transducer which can produce a direct digital output. This device could then dispense with the intricacies of both signal conditioning and analogue to digital conversion. The fact remains however that there are few devices available which can easily produce a digitial output in response to a physical entity; the absolute digital shaft encoder is probably the only transducer which satisfies the specification in its fullest sense.

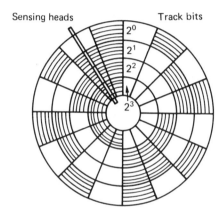

Sensing heads   Track bits

2^0
2^1
2^2
2^3

Figure 2.1   *Digital shaft encoder (4-bit binary code)*

Figure 2.1 illustrates the basic features to be found on a digital shaft encoder. The device consists of a sectioned disc which is mounted onto a rotating element. The disc is divided into a number of tracks and sectors such that on rotation, a different pattern is generated for each resolvable position. The patterns may be detected either optically, magnetically or by direct contact. The resolution achievable is related to the number of tracks on the disc and may be improved with the inclusion of more tracks. For the example given in figure 2.1, there are four tracks which can allow a resolution to 16 angular positions, i.e. rotation through angular segments of 22.5 degrees. With 12 tracks, the resolution can be improved to 2^{12}, i.e. 4096 separately identifiable segments, or ±0.09 degrees of rotation. To avoid inaccuracies due to misalignment of the detectors, the Gray code, see section 4.2, is normally used in preference to the binary code which serves for illustration purposes in figure 2.1. The Gray code is configured such that only one bit is altered for each consecutive resolvable position. Almost all other transducers are analogue in nature, having an infinitely variable output, as opposed to the discretely variable output of the digital transducer.

2.2 MEASUREMENT APPLICATIONS

(i) Rotational Speed

It is often necessary to measure the speed of some rotating piece of machinery and this particular measurement affords great flexibility in the choice of transducer for the said purpose. The digital shaft encoder, previously discussed, lends itself well to this application and complete units are readily available, such as RS 631-632. The RS Components unit comprises an incremental optical shaft encoder which generates 100 pulses per revolution, with an output of 1 volt peak-to-peak and a maximum operating frequency of 20 kHz; equivalent to 12,000 rev/min.

A less expensive alternative is possible using a magnetic pick-up, such as RS 304-166, which responds to the movement of ferrous parts past its pole piece. Figure 2.2 shows the basic configuration which requires the transducer to be in close proximity, 2.5 mm maximum, to a toothed wheel.

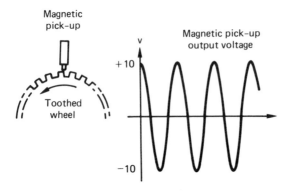

Figure 2.2   Magnetic pick-up speed sensor

A tachometer IC, LM2917N-8, such as RS 302-047, is also available to interface directly to the pick-up. The tachometer is actually a frequency-to-voltage converter where an input pulse rate, the pick-up output signal, is converted to an analogue voltage, or current, which is in direct proportion to the pulse frequency. The pick-up is a passive element requiring no external supply and a schematic diagram of the tachometer is shown in figure 2.3.

The tachometer output is reasonably linear over an input frequency range

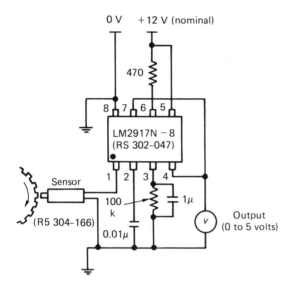

Figure 2.3   Tachometer IC

from about 50 Hz to 800 Hz. The calibration constant however is dependent on the tolerances of the external resistive and capacitive elements which make up the circuit, and the system generally requires calibration for accuracy.

A digital version of the same transducer, RS 304-172, having a square pulse output which is compatible with most logic systems, is also available.

A third alternative for rotational speed measurement involves the use of various opto-electrical devices. These basically incorporate an infra-red emitting device, usually a light emitting diode (LED) and a photo-sensitive detector. One particular unit, RS 307-913, is depicted in figure 2.4.

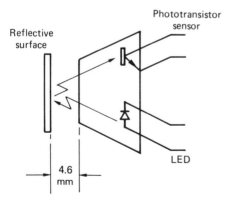

Figure 2.4   Opto-electrical sensor

The passage of a reflective surface, e.g. a toothed cog in a rotating disc, can easily be sensed and coupled with accurate timing, the rotational speed being subsequently determined. Slotted opto-switches, such as RS 304-560, where the toothed disc passes between the emitter and detector, are particularly well suited to this purpose.

A suitable pulse-counting program in BBC BASIC using the TIME function to count the pulse rate input to the least significant bit on the user port, see section 7.2, is as follows:

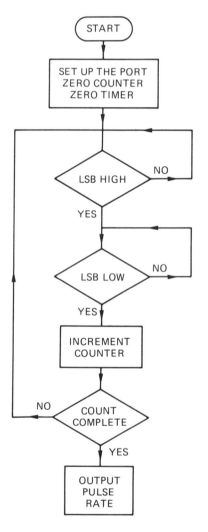

```
 10 ?&FE62=0
 20 INPUT"No.of pulses to be counted="N%
 30 K%=0:TIME=0
 40 REPEAT:UNTIL (?&FE60 AND 1)=1
 50 REPEAT:UNTIL (?&FE60 AND 1)=0
 60 K%=K%+1
 70 IF K%<>N% THEN 40
 80 T=TIME
 90 F=(N%/T)*100
100 PRINT"Frequency = ";F;" Hz"
110 GOTO 30
```

By measuring the time for at least 100 pulses, the program can quite accurately measure the frequency of a pulse train up to about 200 Hz. This is equivalent to a rotational speed of 6000 rev/min with two pulses/rev.

For higher frequencies, a machine code routine using the pulse counting facility available with the 6522 VIA timers might be used. See section 5.3 for further details.

Inductive, capacitive or opto proximity switches can also be utilised for speed measurement. These devices, however, all have considerably lower sensing rates which tends to limit the maximum rotational speed which can be measured. Proximity switches generally are quite expensive and since they are not specifically intended for speed measurement, they cannot be especially recommended for this purpose.

(ii) Temperature

Temperature measurement is commonly required in system monitoring applications and general experimentation. For computer-based systems, thermoelectric devices are employed to generate the necessary output in voltage form. The physics of the thermoelectric effect are complex and even now are not precisely understood. The effect is a consequence of the fact that the number of free electrons in a piece of metal depends both on the temperature and the composition of the metal. For practical purposes, it is sufficient to be aware that when two dissimilar metals are connected together at their ends, then a potential difference will exist if the two ends are held at different temperatures. This physical phenomenon can be used as a means of temperature measurement in the manner illustrated in figure 2.5.

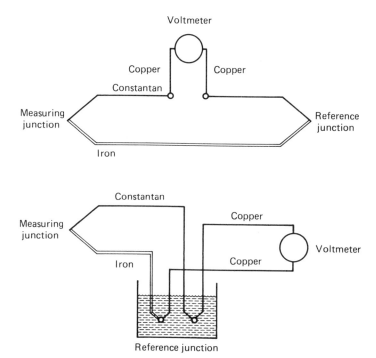

Figure 2.5 Basic thermocouple circuits

Note that in both cases a third metal, copper, has been introduced into the circuit in wiring up to the voltmeter. This has no effect on the thermocouple characteristic provided that the temperatures at the two terminal connections are ostensibly the same.

Since the thermocouple is a differential rather than an absolute measuring device, a known reference temperature is required at one of the junctions if the temperature at the other junction is to be inferred from the output voltage. The most common reference temperature readily available is that of melting ice at standard, or near standard, atmospheric pressure conditions, i.e. 1.01325 bar. This represents zero degrees Centigrade and a range of specially selected materials have been characterised and used as standard thermocouple combinations. The reference temperature need not be 0°C but can be any other known temperature. The reduced potential difference which results with the reference temperature greater than 0°C, can be compensated by adding on the potential which normally exists between 0°C and the reference temperature which is being used. This introduces an error however since the relationship between voltage and temperature although very near, is not quite linear.

A common combination of materials used in general thermocouple applications are nickel-chromium (Chromel) and nickel-aluminium (Alumel). These give the standardised 'K' calibration, approximately 4 mV per 100°C rise above 0°C.

With the reference junction at 0°C, the output voltage at 100°C is only 4.095 mV. While this is perfectly adequate for manual measurements, the output signal requires to be amplified if it is to be processed by microcomputer hardware. A useful and inexpensive self-contained thermocouple amplifier, AD595, such as RS 301-779, is shown in figure 2.6.

The amplifier, housed on a 14 pin monolithic chip, is used in conjunction with type K thermocouples and produces an output of 10 mV/°C, measuring

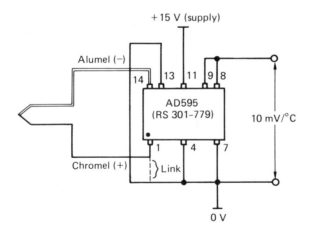

Figure 2.6   *Type K thermocouple amplifier*

from 0°C up to 1100°C. The chip incorporates an ice point compensation circuit which adds a level to the thermocouple voltage which is in direct proportion to the potential difference which exists between 0°C and the chip operating temperature. The chip operating temperature is nominally room temperature since the device has negligible self-heating characteristics.

A simple means of boosting the sensitivity of the basic thermocouple signal is to use a series arrangement called a 'thermopile', figure 2.7.

With the three junction series arrangement as shown in figure 2.7, the output is three times that obtained with a single thermocouple. It is important in this layout that the junctions are electrically insulated from one another and that the hot and cold pile junction temperatures are reasonably uniform.

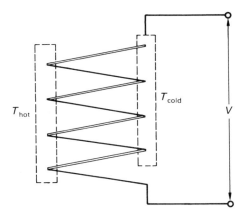

Figure 2.7   *Three-junction thermopile*

A cost-effective, viable alternative to thermocouple circuitry are thermally sensitive resistors, or 'thermistors' as they are generally termed. These are semiconductors with a variable, but very high negative temperature coefficient of resistance such that their resistance is a decreasing function of temperature, see figure 2.8.

The relation between resistance and temperature is given by:

$$R = R_0\, e^{-(b/T)}$$

where R_0 is the thermistor resistance at 0°C
        T is the temperature in degrees Kelvin
and b is the temperature coefficient of resistance.

Thermistors are extremely sensitive and have fast response times, but because of their non-linear character and a general absence of standards, they are not used extensively in industry for temperature measurement. Thermistors with a positive temperature coefficient of resistance are also available and have applica-

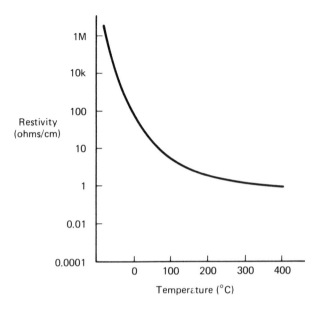

Figure 2.8   *Typical thermistor resistance characteristic*

tion in over-temperature protection circuits and temperature switches. Bead thermistors, e.g. GL16, such as RS 151-013, are available as temperature sensors and are usually incorporated into a bridge circuit, see section 2.3, to provide a suitable output signal.

Temperature transducers with a linear characteristic are also available in integrated circuit form, e.g. RS590kH such as RS 308-809. This transducer acts as a constant current regulator, passing 1 mA per °C and is particularly suited to remote sensing applications where long transmission line resistance and noise may easily degrade the signal from a standard K-type thermocouple. A suitable amplification circuit for this IC is given in figure 2.9.

The two variable resistors shown in the circuit are used to adjust the output voltage and can be arranged to give a calibration of 0–100°C, corresponding to 0–5 volts. This component denoted '741' is a general purpose d.c. amplifier, see section 2.4.

If high accuracy and long term stability are required in the temperature measuring system, then 'platinum-resistance thermometry' may be a more viable alternative. Platinum is stable, corrosion and oxidation resistant, has a high degree of purity and an almost linear resistance characteristic with temperature, about 0.4% per 100°C increment. The near-linear measuring range extends between −200°C and +850°C and, as with thermistors, the platinum-resistance sensor is incorporated into a bridge circuit in practical measurement systems. A typical insertion type, platinum-resistance sensor is RS 158-985.

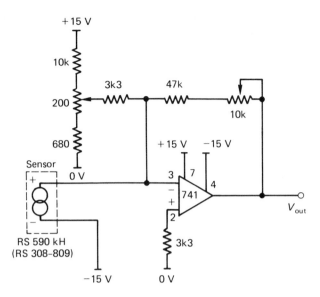

Figure 2.9 *Temperature transducer*

In all of the temperature sensing devices considered so far, the maximum limiting temperature is of the order of 1000°C. The S-type thermocouple, platinum/10% rhodium, platinum, can operate up to 1700°C, but for higher temperatures radiation thermometry is the only workable alternative. Radiation devices are usually optical in nature and require an operator to use visual judgement in the temperature measurement. Obviously the response of the human eye cannot be interfaced to a microcomputer and although pyro-electric detectors are available, they have not yet been developed to the stage where they may operate as a high temperature sensor. Various types of photo-electric devices have been successfully developed however, particularly the photo-voltiac cell or 'solar-cell'. The photo-voltiac cell generates an e.m.f. which is dependent on the intensity of the incident radiation and sensors operating on this principle are used in the steel-making industry. These devices are specialised however and are beyond the scope of this text.

As a precautionary note it must be stressed that temperature measurement, especially at more extreme temperature levels, is not easy and may well be subject to considerable error. The errors may arise because all temperature measurements involve some heat transfer process. A temperature sensor may gain, or lose, heat through the mechanisms of conduction, convection, radiation, or any combination of these. If the heat transfer rates to, or from, the sensor are significant, then the sensor output may be far removed from that which would be measured in the absence of such effects. Other sources of error may be derived from self-heating effects in resistance thermometry, thermoelectric potentials, sensor degradation, transient response times and resistance effects in connection wiring. The latter error source can generally be minimised through using special bridge circuits, but the potential user must, at least, be well aware of the possible sources of error when making important temperature measurements.

(iii) Displacement

(a) Variable Potential Divider

The variable potential divider may be used to convert displacement into voltage, as shown schematically in figure 2.10.

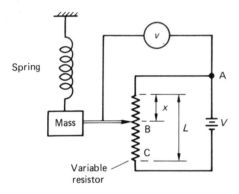

Figure 2.10   *Variable potential divider*

By measuring the voltage between points A and B, a linear relationship between displacement x and voltage v is obtained:

$$x = v(L/V) \tag{2.1}$$

The maximum voltage, $v = V$, occurs when the sliding contact at B reaches the maximum displacement at $x = L$.

As a displacement transducer, however, the variable potential divider has a number of limitations. Equation (2.1) is only strictly applicable if the internal resistance of the voltmeter used to measure v can be considered infinite. This can be illustrated by including the voltmeters internal resistance in the circuit, figure 2.11.

There are two resistances in parallel with each other and using Ohm's law for parallel resistors:

$$\frac{1}{R_{total}} = \frac{1}{R_i} + \frac{1}{R_x} \tag{2.2}$$

where R_x is the resistance along the length of resistor denoted by x.

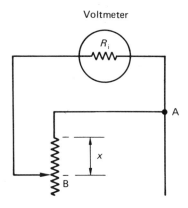

Figure 2.11

The total resistance is therefore given by:

$$R_{total} = \frac{R_i \, R_x}{R_i + R_x} = \frac{R_x}{1 + (R_x/R_i)} \qquad (2.3)$$

It is apparent that R_{total} will approach the value of R_x only when R_i tends to infinity. This is generally referred to as a 'loading error' which causes the net resistance between A and B to be reduced because of the additional circuit through the voltmeter.

Further limitations are associated with the sliding contact which is a source of friction and could also be prone to wear problems. A range of precision linear displacement transducers is available however, and a summary of their main specifications is included in table 2.1.

Table 2.1   Precision linear displacement transducers

RS no.	Stroke length (mm)	Mechanical life (strokes)
RS 317-780	10	5×10^6
RS 159-180	50	10^7
RS 158-890	100	10^7
RS 159-196	200	10^7

(b)   Linear Variable Differential Transformer (LVDT)

A more commonly encountered displacement transducer is the linear variable differential transformer, see figure 2.12.

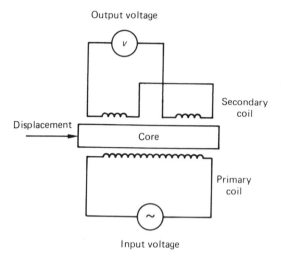

Figure 2.12   *Linear variable differential transformer*

These transducers incorporate an iron core which can move within a primary coil and two overwound pick-up, or secondary, coils. The relative position of the core within the coils produces a proportional a.c. output voltage which can be converted to a proportional d.c. voltage via rectification and smoothing circuits. LVDT transducers have no friction and wear problems since there are no sliding contacts involved. They have infinite resolution, are highly linear and accurate but they are also relatively expensive, e.g. DC50, such as RS 646-511.

There is in addition a wide range of other displacement transducers which utilise either inductive or capacitive effects. These are generally restricted to the measurement of very small displacements, e.g. 0.1 mm.

(iv) Pressure

The measurement of pressure is an indirect measurement where, not uncommonly, it is displacement which is actually measured and the resulting signal is related to the pressure which caused the displacement. The conventional pressure transducer consists of an elastic element which can deflect under an applied pressure. The varied means of sensing the deflection give rise to an equally varied range of pressure transducer, some of which are depicted schematically in figure 2.13.

Many other mechanical configurations are possible but the common factor lies in the ultimate conversion of displacement to d.c. voltage output.

Pressure transducers based on the piezoelectric effect belong in a separate class. A piezoelectric crystal, e.g. quartz, produces an electrostatic charge when subjected to a mechanical strain. Since the strain is the result of an applied force, or pressure, then the electrostatic charge forms the basis of the force, or pressure,

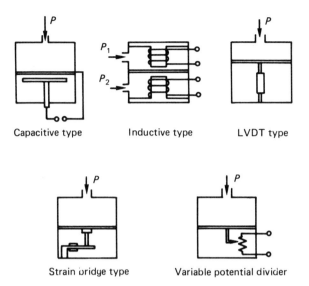

Figure 2.13   Various pressure transducers

measurement. A difficulty exists however since the charge is manifested as a consequence of the application of the mechanical load. If the charge were removed, then no new charge would be generated until the loading state again changes. This feature makes piezoelectric pressure transducers unsuitable for static measurements, but eminently suitable for dynamic pressure monitoring. Piezoelectric transducers are commonly used for example, to measure the time variation of pressure within the cylinders of internal combustion engines.

Measurement of the charge however, without dissipating it, is problematic such that piezoelectric transducers require sophisticated signal conditioning, incorporating a charge amplifier. This, in turn, results in very expensive measurement systems.

A summary of the main pressure transducer characteristics is given in table 2.2.

Table 2.2   Pressure transducer characteristics

Type	Typical pressure range (kN/m^2)	Response time (s)
Capacitive	0.01–200	0.01
Inductive	0.1 –200	0.01
Strain bridge	20 –25000	0.001
Piezoelectric	0.1 –15000	0.0000033

The figures given in the table are typical, but not restrictive as special transducers are available, in all types, which are designed to operate outwith the stated ranges.

Prices for pressure transducers vary considerably because of the type of pressure connections required and the nature of the fluid to be monitored. The strain bridge type is generally the cheapest. Capacitive and inductive type pressure transducers, complete with signal conditioning circuitry, are more expensive but the piezoelectric-based pressure transducers are the most costly. A recent inexpensive newcomer, however, is the piezo-resistive type of pressure transducer, such as RS 303-373.

(v) Flow Measurement

The measurement of flow covers a wide range of different techniques and the subject often forms the basis of a textbook in its own right. For microcomputer-based data-acquisition systems the only flow metering devices of any consequence are those which either have a direct electrical output, or those which, with an additional transducer, can produce an electrical signal related to the flow rate.

(a) Differential Pressure Flowmeters

These include venturi-meters, orifice-meters or indeed any device which presents a constriction in the flow path and causes the fluid to accelerate with an attendant drop in pressure. The operating principles of these types of meters are based on two fundamental laws of incompressible fluid dynamics. These are the steady flow energy equation, or Bernoulli equation, and the equation of mass continuity. In applying these laws to any types of flow constriction an equation for the flow rate emerges in the general form:

$$\text{Flow rate} = C_d \, A_1 \sqrt{\left(\dfrac{\dfrac{2}{\rho}(p_1 - p_2)}{\left[\left(\dfrac{A_1}{A_2} \right)^2 - 1 \right]} \right)} \tag{2.4}$$

where C_d is a discharge coefficient: a 'constant' to accommodate frictional, jet contraction and other losses in the device

A_1 is the upstream pipe cross-sectional area

A_2 is the constriction cross-sectional area

ρ is the fluid density, assumed constant

p_1 is the upstream fluid static pressure

p_2 is the downstream fluid static pressure.

The equation can be reduced to the form:

$$\text{Flow rate} = K \sqrt{(\Delta p)} \tag{2.5}$$

where K is, in general, an experimentally determined constant.

and Δp is the measured pressure differential across the meter.

The measurement of Δp could be performed with a suitable pressure trans-ducer and provided that K is a well behaved 'constant', the flow rate can be accurately calibrated against the voltage output of the transducer.

If the fluid is dry air at low pressure, it is usually necessary to measure the air temperature and the barometric pressure in addition to the meter differential pressure.

Figure 2.14 shows three typical differential pressure flowmeters used exten-sively in industry.

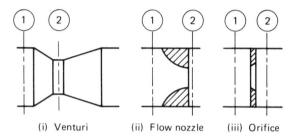

Figure 2.14   *Differential pressure flowmeters*

(b)   Positive Displacement Meters

Positive displacement devices are designed such that a metered fluid repeatedly passes through a known fixed volume. The rotary-vane flowmeter, figure 2.15, typifies the general principle in that for each revolution of the rotor, a fixed quantity of fluid is passed through from inlet to outlet.

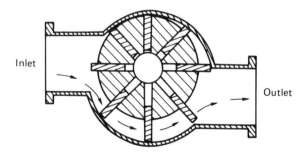

Figure 2.15   *Rotary-type flowmeter*

Measurement of the rotor speed gives a proportional measure of the volumetric flow rate. The lobed-rotor meter, figure 2.16, also comes under this category, which is generally accurate to about ±0.5%.

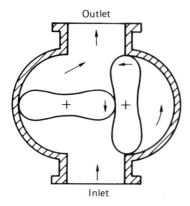

Figure 2.16   *Lobed-rotor flowmeter*

(c)   Turbine Meters

Turbine meters incorporate some multi-vaned rotor which is driven by the metered fluid, figure 2.17.

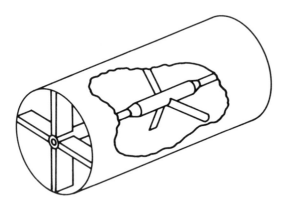

Figure 2.17   *Turbine-type flowmeter*

Provided mechanical friction is minimal, then the rotational speed of the turbine equates in direct proportion to the volumetric flow rate. The rotational speed is often measured using a magnetic pick-up, section 2.2, which produces a pulsed output in response to the passage of a rotor blade. A frequency-to-voltage converter can be interfaced to the pick-up to provide a d.c. output voltage in direct proportion to speed and thus also to flow rate.

Inexpensive rotary flowmeters, for use with liquids, are available for low to moderate flow rates with the general specification as given in table 2.3.

Table 2.3 Rotary flowmeters

RS no.	Flow rate (litres/hour)	Output frequency
RS 304-431	3–100	24 Hz at 10 l/h
		52 Hz at 20 l/h
RS 630-746	10–500	31 Hz at 30 l/h
		375 Hz at 300 l/h

To complete the measurement system, the tachometer IC, figure 2.3, may conveniently be connected as shown in figure 2.18 to function as a frequency-to-voltage converter, with an output ranging between 0 to 5 volts d.c.

Higher quality turbine flowmeters are very expensive and complete systems usually incorporate the turbine, the speed sensor and the signal conditioning circuitry in a single unit.

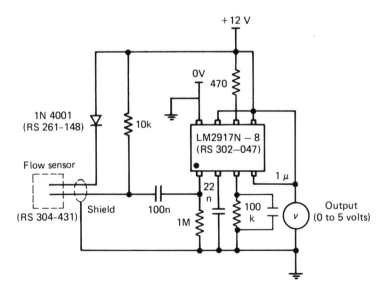

Figure 2.18 Flowmeter signal conditioner

(d) Drag Plate Devices

Flowmeters based on the principles of fluid dynamic drag are typified by the example illustrated in figure 2.19.

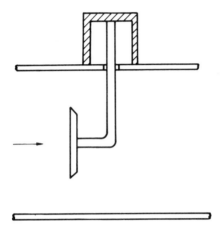

Figure 2.19 *Simple drag-plate flowmeter*

A simple target plate is exposed to the oncoming flow which causes a fluid dynamic drag to be exerted on the plate. The drag force is given by the relation:

$$\text{Drag force} = C_d \, A \tfrac{1}{2} \rho U^2 \tag{2.6}$$

where C_d is a drag coefficient

  A is the projected cross-sectional area of the plate

and U is the mean fluid velocity approaching the plate.

The drag force exerted on the plate causes the supporting arm to flex and the displacement can be measured via a strain bridge, or other suitable transducer. The measured displacement can then be related to the flow rate through a calibration. This sort of flowmeter is dependent on having a near constant value of drag coefficient, which can be approximately obtained if the target plate is in the form of a flat, sharp edged lamina. Drag plate flowmeters require calibration against some other standard for accuracy. They often form the basis of many a 'do-it-yourself' flowmeter.

(e) Vortex Shedding Devices

When a 'bluff' body is exposed to an oncoming fluid, the fluid particles in the proximity of the surface cannot follow the surface contours and they break away from the surface and form periodic vortices, or eddies. The vortices are shed alternatively from the upper and lower extremities of the bluff shaped body, figure 2.20.

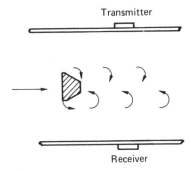

Figure 2.20 Vortex shedding flowmeter

Over a considerable range, for a particular shape of body, the periodicity of the vortex shedding is in direct proportion to the fluid velocity and hence also to the flow rate. A measurement of the shedding frequency can therefore be translated to flow rate. The means of sensing the shedding frequency are varied and include among others, pressure variation using a piezoelectric pressure transducer, resistance variation using a heated thermistor and interference of an ultrasonic beam transmitted across the wake generated behind the body. Vortex shedding flowmeters have no moving parts and can be used with any fluid; the output signal in general is quite noisy however and the linear operating range may be restrictive. A frequency-to-voltage converter normally completes the instrumentation set-up.

Many other commercial flowmeters are available, ranging in complexity and price, and table 2.4 presents a somewhat condensed summary of some other common types.

With the exception of the variable area flowmeters, or 'rotameters' as they are often called, all the other flow metering devices are very expensive and the two anemometry systems, in addition, are generally restricted to research labora-

Table 2.4 Flow metering instrumentation

Type	*Operating principle*	*Application*
Electromagnetic	Induced e.m.f.	Industrial
Ultra-sonic	Doppler shift frequency	Industrial
Hot wire/film anemometry	Resistive bridge	Applied research
Laser–Doppler anemometry	Doppler shift frequency	Applied research
Variable area	Drag balance	Industrial

tory investigations where the measurement of local fluid velocities are required. Robust, general purpose instruments, based on the hot wire anemometer, are available however at reasonable cost. These are particularly suited to low velocity measurements and have application in fume cupboards, environmental chambers and ventilation ducts.

The measurement of many other physical variables is derived from the measurement of a related effect that the physical quantity has on a well defined system. Displacement, for example, is particularly versatile in this respect since it could be a consequence of the effects of pressure, force, acceleration, torque, stress, velocity, vibration or seismic activity. A measured displacement therefore could represent any of the above listed parameters with the appropriate conditioning and calibration of the basic sensor mechanism.

The measurements discussed so far, in fact embrace a wide spectrum of all the measurements which are made in the experimental and industrial contexts. Other measurements are of a more specialised nature which is beyond the scope of the present text. If the reader has a specific application where some of the less common measurements are to be made, then further details can be obtained from the sources of information quoted in the references at the end of this chapter.

2.3 STRAIN GAUGE APPLICATIONS

In transducer technology, 'bridge' networks figure prominently in the sensor circuitry to the extent that they justify some extended consideration. The most prevalent bridge circuits involve strain gauge resistive elements although capacitive and inductive bridges are also utilised in the measurement of steady-state and transient phenomena. The basic Wheatstone bridge is shown in figure 2.21.

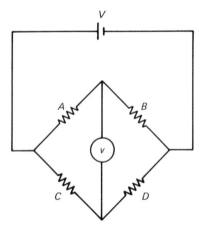

Figure 2.21 *Resistive Wheatstone bridge*

The bridge consists of four resistors, A, B, C and D, a voltage source and usually a high impedance instrument to record the bridge output voltage.

When the bridge is 'balanced' then no current is drawn through the voltmeter. From Ohm's law it can be shown that for balance the ratios (A/C) and (B/D) are equal. Thus, having C a variable resistor, A may be determined if C, B and D are all known. The result is independent of the supply voltage.

In most cases the sensing element forms one arm of the bridge while the other three resistors are fixed, but equal to the sensing element resistance under particular reference conditions. These bridges normally operate in the unbalanced mode where the voltmeter reading is used to quantify the changing conditions at the sensing element; the result of changes in the physical variable being measured.

The additional variable resistor S, shown in figure 2.22, is often incorporated into the circuit as a current limiting device to prevent self-heating and subsequent variation in the resistance of the individual bridge elements. The magnitude of S will influence the sensitivity of the circuit.

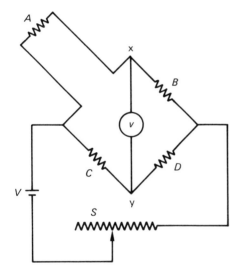

Figure 2 22   Unbalanced Wheatstone bridge

Removing S from the circuit and assuming that the voltmeter has an infinite internal resistance, the voltage developed across the meter can be shown to be approximately linear. Initially it is assumed that the resistance of all the bridge elements is R and that the bridge is in balance. The sensing element, A, is then subjected to a condition which causes its resistance to change to $(R + \Delta R)$ and thereby causes a voltage v to be generated across the voltmeter. Since the voltmeter has infinite resistance then it draws no current. The current through C and

D must be constant and equal to $(V/2R)$, where V is the voltage applied across the bridge. The current through A and B is $(V/(2R + \Delta R))$.

The potential available at x is therefore $V - \dfrac{V(R + \Delta R)}{(2R + \Delta R)}$

Similarly, the potential at y is $V - \dfrac{VR}{2R} = \dfrac{V}{2}$

The potential difference, $(x - y)$, is therefore:

$$v = \left\{ V - \frac{V(R + \Delta R)}{(2R + \Delta R)} \right\} - \frac{V}{2}$$

$$= \frac{V}{2} - \frac{V(R + \Delta R)}{(2R + \Delta R)}$$

$$= \frac{V}{2} \left\{ 1 - \frac{(R + \Delta R)}{\left(R + \dfrac{\Delta R}{2} \right)} \right\}$$

$$= \frac{V}{2} \left\{ \frac{\left(R + \dfrac{\Delta R}{2} \right) - (R + \Delta R)}{\left(R + \dfrac{\Delta R}{2} \right)} \right\}$$

$$= \frac{V}{2} \left\{ \frac{- \Delta R/2}{(R + \Delta R/2)} \right\}$$

$$= - \frac{V}{4} \times \frac{\Delta R}{R} \left\{ \frac{1}{1 + \dfrac{1}{2} \times \dfrac{\Delta R}{R}} \right\}$$

The final relation is non-linear. Nonetheless, if $R \gg \Delta R$, then:

$$v \simeq \frac{V}{4} \frac{\Delta R}{R}$$

which gives an approximately linear dependence of v on ΔR.

If the voltage recording instrument has a low internal resistance, e.g. a sensitive, centre-zero galvanometer, then the above analysis does not strictly apply. Using Thévenin's theorem however, it can be shown that a near linear dependence still applies in these cases, provided that ΔR remains small.

The advantage of the unbalanced bridge lies with the direct output voltage which is generated. The unbalanced bridge however lacks the intrinsic accuracy of the null-reading balanced bridge and the output is dependent on the supply voltage, V, which must be kept constant.

Wheatstone bridges may also be energised with alternating current, in which they become known as impedance bridges. The same principles apply, although for equal ratios of bridge impedances only the magnitude of the voltages across the measuring points will be equal. The potentials may still be out of phase and an a.c. sensitive instrument would indicate imbalance. Because of the added complication associated with a.c. bridges, they are usually calibrated empirically and used only for very special applications. Alternating current bridges can incorporate any type of impedance element including both capacitive and inductive devices.

The resistive element Wheatstone bridge provides the circuitry to be used in conjunction with a variable resistance as the basic transducer sensor. The sensor element may be a thermistor in a temperature measuring application, or a strain gauge in a displacement measuring system.

The principle of the strain gauge is that a conductor, of initially uniform dimensions, will alter its proportions when subjected to an applied stress. The resistance of the conductor is also dependent on the physical dimensions such that the applied stress, in addition to deforming the conductor shape, will manifest a change in its total resistance.

The resistance of the conductor is given by:

$$R = \beta \, (L/A) \tag{2.7}$$

where β is the resistivity of the material

L is the conductor length

and $\quad A$ is the conductor cross-sectional area.

For a cylindrical conductor of diameter D, the cross-sectional area is given by $A = \pi D^2 /4$.

Differentiating equation (2.7) results in:

$$\frac{dR}{R} = \frac{d\beta}{\beta} + \frac{dL}{L} - \frac{dA}{A}$$

and $\quad dA = \frac{\pi}{2} D \, dD$

$$\therefore \quad \frac{dA}{A} = \frac{\frac{\pi}{2} D \, dD}{\pi D^2 /4} = 2 \frac{dD}{D}$$

Dividing through by (dL/L) gives:

$$\frac{dR}{R} \frac{L}{dL} = \frac{d\beta}{\beta} \frac{L}{dL} + 1 - \frac{2dD}{D} \frac{L}{dL} \tag{2.8}$$

(dL/L) is the ratio of the change in length of the conductor to the original length and is the axial strain, ϵ_a.

Similarly, (dD/D) is the transverse strain, ϵ_t.

and $\left(\dfrac{dD}{D}\dfrac{L}{dL}\right)$, that is $\left(\dfrac{\epsilon_t}{\epsilon_a}\right)$, is Poisson's ratio, ν, for the material. Substituting these parameters into equation (2.8) gives;

$$\left(\frac{dR}{R}\frac{L}{dL}\right) = \frac{d\beta}{\beta}\frac{1}{\epsilon_a} + 1 + 2\nu \tag{2.9}$$

The subject of equation (2.9), $\left(\dfrac{dR}{R}\dfrac{L}{dL}\right)$, is referred to as the gauge factor, G, and is specified by the manufacturer. The constancy of the right-hand side of equation (2.9) will determine the suitability of any material for application as a strain gauge.

Since $G = \dfrac{dR}{R}\dfrac{L}{dL} = \dfrac{dR}{R}\dfrac{1}{\epsilon_a}$, then:

$$dR = G R \epsilon_a \tag{2.10}$$

With G and R both known, the axial strain is in direct proportion to the change in resistance, dR. It is apparent from equation (2.10) that a high value of G would be advantageous in producing a high change in resistance for any input axial strain. The more popular strain gauge materials have gauge factors of the order of about 2, although some semi-conductor devices have gauge factors in excess of 100. The commonest form of strain gauge is the bonded foil type, figure 2.23.

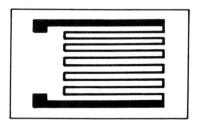

Figure 2.23   Bonded foil type strain gauge

Foil-type strain gauges must be rigidly fixed to the strained member and electrically insulated from it. Commercial gauges have nominal unstrained resistances in the range 50 to 1000 Ω and can respond to a change in strain, typically of the order of 10^{-6} (or 1 micro-strain). With a gauge factor of 2 and an unstrained resistance of say 120 Ω, equation (2.10) gives the change in resistance, dR equal to 0.24 mΩ per micro-strain. Because the change in resistance is so small, strain gauge measurements are inherently prone to error through temperature effects and it therefore becomes essential to compensate for temperature variation in any practical measuring system. Figure 2.24 illustrates the simplest method of temperature compensation by the inclusion of a second, identical strain gauge as the balancing resistance in a Wheatstone bridge.

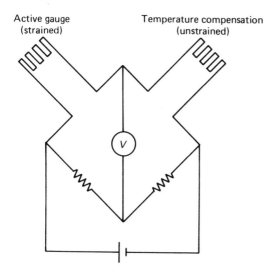

Figure 2.24   Strain gauge bridge with temperature compensation

The second strain gauge is at all times subject to the same ambient temperature as the active, or measuring gauge, but it is attached to a separate, unstrained piece of similar material to that of the active gauge. This effectively counterbalances any resistance changes attributable to temperature variations and the bridge voltage output will then be a consequence only of applied strain on the active gauge. As an alternative, self temperature-compensating gauges may be used in situations where compensation with a 'dummy' gauge is impracticable. Strain gauges are available, such as RS 632-168, with self temperature compensation for either steel or aluminium.

The sensitivity of the basic circuit may be increased if it is known that the flexural member is subject to equal and opposite strains on its opposite surfaces. Figure 2.25 indicates a possible application for a force transducer in the form of a simple cantilever.

Analysis of the circuit in a similar manner to that given previously results in:

$$v = \frac{V}{4} \frac{2\Delta R}{R}$$

It can be seen that the output, v, has been doubled and that the sensitivity of the circuit has been increased by a factor of 2. By using active gauges on all four arms of the bridge, the sensitivity can again be doubled with:

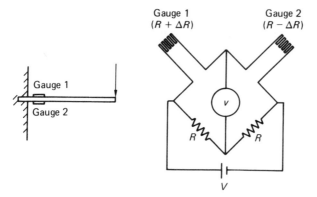

Figure 2.25 *Simple force transducer*

$$v = V \frac{\Delta R}{R}$$

In either case, temperature compensation is automatic and the four active arm strain gauge bridge often forms the basis of a secondary transducer for the indirect measurement of pressure, force, torque, or any similar parameter which ultimately causes deflection in an elastic member. If the four-arm bridge is to be used in this secondary capacity, then the actual strain is of no particular interest and the gauge factor becomes unimportant. The output voltage will then be calibrated against the primary variable and the intermediate processes are of no consequence.

For applications where the strain is the primary variable then a bridge zero-adjustment becomes essential since it is virtually impossible to ensure equality of the resistances within the tolerances demanded by the sensitivity of the circuit. Other problems associated with the basic signal are electronic noise pick-up, zero-drift, non-linearity and the necessity of a precision amplifier to produce an output suitable for interfacing to a microcomputer. Notwithstanding the difficulties associated with strain gauge bridge circuits a custom built amplifier, SGA100, such as RS 308-815, is available along with a suitable printed circuit board, RS 435-692, at very reasonable cost to make an amplifier/decoder for resistive bridge circuits. This circuit may be used in conjunction with a strain gauge bridge to measure strain, or to function in the secondary capacity as, say, a load cell. In the latter role, the load bearing unit may be designed to suit the requirements of the loading system to be measured. An appropriate 'cell' for the measurement of compressive loads might take the form depicted in figure 2.26.

The strain gauges are arranged at 90 degrees to each other in 'rosette' fashion and four gauges are attached to the flexural member. This layout helps to augment the output signal since all of the gauges are active. The two gauges aligned in the direction of the applied load tend to contract, while the two

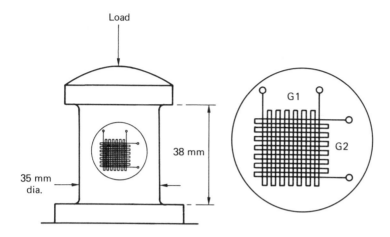

Figure 2.26 *Simple load cell for compressive load measurement*

gauges aligned at 90 degrees to the load tend to elongate since the circumference of the stressed member increases with applied load.

To provide temperature compensation in the connection wiring, the so-called 'five-wire' circuit shown in figure 2.27 may be used.

Amplification of the bridge output is achieved with the amplifier, SGA100, and the additional circuitry given in figure 2.28.

Although the circuit given is quite complex, its construction is greatly facilitated with the ready made printed circuit board which comes supplied with a list of essential components and RS stock reference numbers.

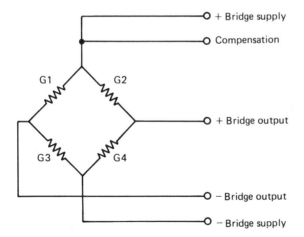

Figure 2.27 *Full bridge (five wire) circuit for a compressive load cell*

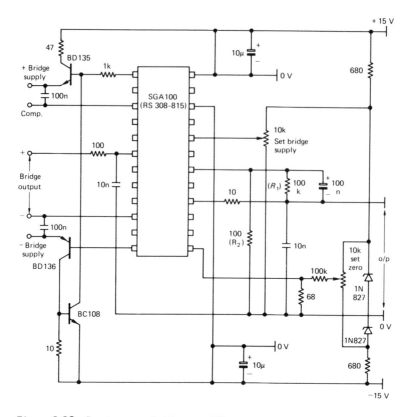

Figure 2.28 *Strain gauge bridge amplifier*

The amplifier gain can be altered and is defined by the relation:

Gain $= 1 + R_1/R_2$

For the values of R_1 and R_2 given in the figure, the resultant gain is approximately 1000. Other values may be set if so desired.

With the load cell manufactured from aluminium bar, a load calibration similar to that given in figure 2.29 might be obtained.

The total cost of the components comprising the system described is fairly minimal. As an alternative for strain measurement, complete systems can be purchased. These may be very expensive however, depending on the number of channels required, sensitivity and accuracy etc.

For further practical details on strain gauge selection and installation, the reader is referred to the *Student Manual* available from Welwyn Strain Measurement, see references.

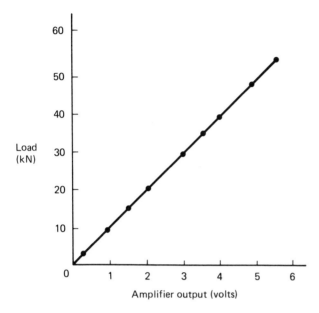

Figure 2.29 *Typical load cell calibration*

2.4 SIGNAL CONDITIONING – THE OPERATIONAL AMPLIFIER

In the three preceding sections it becomes apparent that although the multitude of transducers which are available can produce the necessary output in voltage form, the output signal available is, in most cases, still unsuitable for direct interfacing to a microcomputer system. Transducer output voltages are often in the milli-volt range and as such they will usually require amplification before a useful signal can be obtained for interfacing purposes. The amplification process can in itself introduce other problematical features including noise, stability, common mode rejection and distortion. Some special purpose amplifiers, in IC form, have already been introduced in this chapter but a more general treatment is now presented since amplification constitutes an important general feature in instrumentation technology.

The term amplifier is in fact a shortened form for the full description, 'voltage amplifier'. Most amplifiers amplify voltage but other types are enountered and these are usually given their full description, for example current amplifier, power amplifier or charge amplifier. The prefix denotes the quantity which is amplified. Amplifiers are further classified as a.c. or d.c. in which the d.c. amplifier will accept a.c. inputs, but the a.c. amplifier will 'block' d.c. inputs. The input may be either single-ended or differential with two active signal lines and the amplifier may be either 'non-inverting' with no sign change, or 'inverting' with a reversal of sign at the output.

The commonest configuration embodies a differential input, single-ended output voltage amplifier with either inverted or non-inverted output. These types are referred to as 'operational amplifiers' and a selection of some typical circuits is depicted in figure 2.30.

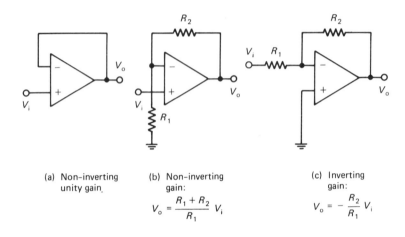

(a) Non-inverting
unity gain.

(b) Non-inverting
gain:
$$V_o = \frac{R_1 + R_2}{R_1} V_i$$

(c) Inverting
gain:
$$V_o = -\frac{R_2}{R_1} V_i$$

Figure 2.30 *Operational amplifier circuits*

The most important characteristics of an operational amplifier are its gain and its bandwidth. The gain is related to the combination of resistors used in the external circuitry of the amplifier and may be as high as 10^6. The gain is constant over a particular range of input signal frequencies, but outside this range the gain tends to fall, or 'attenuate', as shown in figure 2.31.

The plot of gain against signal frequency is called a 'Bode diagram' and the frequency range over which the gain remains above $(1/\sqrt{2})$ of the constant linear value is called the bandwidth. The gain is usually expressed as a logarithmic ratio of the power transmitted between input and output. Since the voltage squared is a measure of the power transmitted then the gain is:

$$\text{gain} = \log_{10}\left(\frac{\text{output power}}{\text{input power}}\right) \text{ bels}$$

$$= 10 \log_{10}\left(\frac{V_o^2}{V_i^2}\right) \text{ deci-bels}$$

$$= 20 \log_{10}\left(\frac{V_o}{V_i}\right) \text{ deci-bels, or db}$$

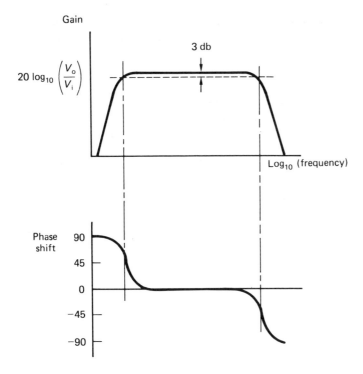

Figure 2.31   Gain and phase shift characteristics for an operational amplifier

Since 20 log $(1/\sqrt{2})$ = -3 db, then the particular frequencies where the gain has been reduced by 3 db are known as the '3 db cut-off frequencies' and these define the bandwidth. The product of gain and bandwidth (GBW) is often quoted by manufacturers as a quality parameter. Other descriptive properties associated with amplifiers are listed in table 2.5.

Table 2.5   Various amplifier characteristics

Property	Description
Common mode rejection ratio	Ability of an amplifier to reject differential input gain variations
Offset voltage	Voltage outputs attributable to input and output voltages which are generated by component variations
Offset drift	Temperature-dependent voltage outputs
Non-linearity	Departures from linear input/output characteristics
Distortion	Frequency-dependent non-linearities

A popular general purpose operational amplifier is UA741CP, such as RS 305-311. Also available is a ready-made printed circuit board for the amplifier, RS 434-065.

Figure 2.9 shows the so-called '741' used in a circuit as a current/voltage converter, in conjunction with a temperature to current converter, RS590kh, to function as an inexpensive temperature transducer.

For high accuracy measurements, where low drift and low noise are also essential, an instrumentation amplifier with a high input impedance and high common mode rejection ratio would normally be required, figure 2.32.

A precision instrumentation amplifier, INA101HP, such as RS 636-227, is suitable for the amplification of signals from strain gauges, thermocouples and other low level differential signals from bridge circuits and other transducers; see application in section 10.3.

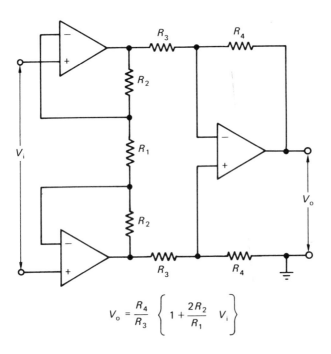

$$V_o = \frac{R_4}{R_3} \left\{ 1 + \frac{2R_2}{R_1} \; V_i \right\}$$

Figure 2.32   Instrumentation amplifier

2.5   FILTERS AND BUFFERS

Noise is inherently present in all physical systems where measurements are performed. The sources of noise are varied and may originate from thermo-electric effects, electro-chemical action, electro-magnetic and electro-static pick-up, self-generated component noise, offset voltages and common earth loops. Whatever the noise source, it is to the experimenters advantage if *a priori* knowledge exists concerning the frequency range in which the signal of interest lies. With this knowledge, it is possible to filter out the signal above or below the frequency range of specific interest. In the study of turbulence in a fluid flow for example, the most significant velocity fluctuations are known generally to lie in the frequency range 0.1 Hz to 2 kHz. It becomes possible therefore to attenuate all of the signal levels above 2 kHz with no loss in the definitive turbulent character of the remaining signal.

Various arrangements may be used for analogue filter circuits, but they are broadly classified as (i) lowpass, (ii) highpass and (iii) bandpass, figure 2.33.

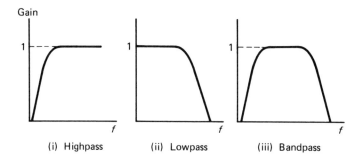

(i) Highpass   (ii) Lowpass   (iii) Bandpass

Figure 2.33 Analogue filter performance curves

A lowpass filter allows the transmission of signals below a particular cut-off frequency. The highpass filter, on the other hand, transmits only that part of the signal above the cut-off frequency while the bandpass filter transmits the signal contained within the range between the upper and lower cut-off frequencies. The cut-off frequency does not represent a discontinuity, but the filter progressively attenuates the signal as the frequency moves away from the cut-off value. The 3 db attenuation level is normally used to specify the filter cut-off frequencies.

The simplest forms of analogue filter incorporate single resistive, capacitive or inductive elements.

(i) Lowpass Filter

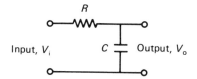

Figure 2.34

The relationship between input and output is given as:

$$\frac{V_o}{V_i} = \frac{\frac{1}{sC}}{R + \frac{1}{sC}}$$

where s is the Laplacian operator $(\partial/\partial t)$.

Hence $V_o (1 + sRC) = V_i$

$$\therefore \ V_o + RC \frac{\partial V_o}{\partial t} = V_i \tag{2.11}$$

The product RC is referred to as the time constant of the circuit and is denoted by τ.

$$\tau = RC = \frac{1}{\omega_c} = \frac{1}{2\pi f_c} \tag{2.12}$$

where f_c is the cut-off frequency in Hz and ω_c is the equivalent circular frequency in radians/sec.

Example

A simple RC circuit is to be used as a lowpass filter. If the output voltage is to be attenuated 3 db at 1 kHz, calculate the required value of the time constant.

From equation (2.12): $\tau = \dfrac{1}{2\pi f_c} = \dfrac{1}{2\pi \times 1000}$

$$= 1.59 \times 10^{-4} \ \text{seconds}$$

The circuit might then be constructed using a 0.1 μF capacitor and a 1.59 kΩ resistor.

(ii) Highpass Filter

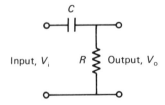

Input, V_i R Output, V_o

Figure 2.35

In this case, $\dfrac{V_o}{V_i} = \dfrac{R}{\dfrac{1}{sC} + R} = \dfrac{sRC}{1 + sRC}$

$\therefore\ V_o\,(1 + sRC) = V_i\,sRC$

$$\therefore\ V_o + RC\,\frac{\partial V_o}{\partial t} = RC\,\frac{\partial V_i}{\partial t} \tag{2.13}$$

Using the appropriate combinations of resistor and capacitor, filtering of the input signal can result in a significant improvement in noise reduction. In the majority of engineering cases, the signal frequencies are quite low in comparison to the ambient noise and lowpass filters are more commonly encountered. Many other forms of filter exist and the reader should consult the references given at the end of this chapter if further details are required on the more complex forms.

With the emergence of fast data-acquisition systems, hard-wired analogue filters can easily be replaced with a digital equivalent. With the input signal available in digital form, the concept of digital filtering is accommodated in software through a finite difference approximation to the governing equation. When dealing with a discretised signal we can denote the 'input' as X and the 'output' as Y. The simple lowpass RC circuit may then be represented as follows:

Input, X ──── Filter ──── Output, Y

with $$Y + \tau\,\frac{dY}{dt} = X \tag{2.14}$$

The cut-off frequency is defined as before, $f_c = \dfrac{1}{2\pi\tau}$.

If the digital sampling interval was Δt, then the sampling frequency is denoted by:

$$f_s = \frac{1}{\Delta t}$$

Using a backward difference approximation, the governing equation (2.11) may be modelled as:

$$Y_i + \frac{\tau}{\Delta t}\,(Y_i - Y_{i-1}) = X_i \tag{2.15}$$

where i denotes the values at time level t
and $(i - 1)$ denotes the values at the previous time level, $(t - \Delta t)$.

Hence $Y_i \left(1 + \dfrac{\tau}{\Delta t} \right) = \dfrac{\tau}{\Delta t} \, Y_{i-1} + X_i$

$$\therefore \; Y_i = \left[\frac{1}{\left(1 + \dfrac{\Delta t}{\tau} \right)} \right] Y_{i-1} + \left[\frac{1}{\left(1 + \dfrac{\tau}{\Delta t} \right)} \right] X_i \qquad (2.16)$$

The time constant, τ, can be set to any desired value but in normal circumstances it would be selected such that the resulting cut-off frequency was no greater than about half that of the sampling frequency. Cut-off to sampling frequency ratios of 1/5 to 1/10 are more common.

Example

An instrumentation signal is measured and converted to digital form at the rate of 7.5 kHz. The frequency range of interest is 0 to 500 Hz and it is intended to attenuate the signal digitally with a 3 db cut-off frequency of 500 Hz. Determine the particular form of the digital filtering algorithm if it is to be based on a simple *RC* analogue filter.

$$f_c = 500 \text{ Hz}$$

$$\therefore \; \tau = \frac{1}{2\pi \times 500} = 0.000\,3183 \text{ seconds}$$

$$\Delta t = \frac{1}{f_s} = \frac{1}{7500} = 0.000\,133 \text{ seconds}$$

From equation (2.16):

$$Y_i = 0.7048 \, Y_{i-1} + 0.2952 \, X_i$$

An advantage that the digital filter has over its analogue counterpart is that the 'signal' may be repeatedly filtered any number of times simply by processing the data repeatedly through the filtering algorithm. Digital filtering takes longer to perform in real time but is easy to modify and adjust.

To illustrate the technique, the following BASIC program uses a pseudo input to assess the performance of the previous digital filtering algorithm. The program includes an optional graphics output applicable to the BBC microcomputer. It is assumed that the input signal was a sine function of amplitude 10 volts peak-to-peak and a frequency of 750 Hz.

With a sampling frequency of 7.5 kHz there would be ten sample points captured for every cycle of the input. The input is generated synthetically in the first FOR–NEXT loop of the program. The cut-off frequency of 500 Hz is retained.

The graphical output is shown in figure 2.36, where the signal is drawn along with the output after one pass through the algorithm and also the output after a further three passes.

```
10 DIM X(50),Y1(50),Y2(50),Y3(50),S(50)
20 REM Generation of pseudo digitised input
30 X(1)=0
40 FOR K=1 TO 49
45 VDU2
50 X(K+1)=5*SIN(RAD(40*K))
60 S(K+1)=X(K+1)
70 NEXT K
80 Y1(1)=X(1)
90 REM First pass through filter
100 FOR I=2 TO 50
110 Y1(I)=0.7048*Y1(I-1)+0.2952*X(I)
120 Y2(I)=Y1(I)
130 Y3(I)=Y1(I)
140 NEXT I
150 N=0
160 REM Additional passes through filter
170 REM Re-assign input as the output from
180 REM the previous pass through the filter
190 FOR K=2 TO 50
200 X(K)=Y3(K)
210 NEXT K
220 N=N+1
230 REM Next pass through filter
240 FOR I=2 TO 50
250 Y1(I)=0.7048*Y1(I-1)+0.2952*X(I)
260 Y3(I)=Y1(I)
270 NEXT I
280 REM Three more passes
290 IF N<2 GOTO 160
300 CLS
310 REM Graphics output
320 MODE1:MOVE 50,400
330 DRAW 1250,400
340 MOVE 50,0:DRAW 50,900
350 MOVE 50,400
360 FOR K=1 TO 50
370 DRAW (K*20)+30,(S(K)*80)+400:NEXT K
380 MOVE 50,400
390 FOR K=1 TO 50
400 DRAW (K*20)+30,(Y2(K)*80)+400:NEXT K
410 MOVE 50,400
420 FOR K=1 TO 50
430 DRAW (K*20)+30,(Y3(K)*80)+400:NEXT K
440 END
```

It can be seen that after a single pass the signal level has been attenuated and there is also a phase shift associated with the process. These are exactly the same effects which would have been observed with the hard-wired analogue filter. With additional passes through the digital filter, the signal is further attenuated and is subject to a cumulative shift in phase with each successive cycle of the numerical process.

The digital filter also exhibits an inherent numerical transient which is related to the processing done on the data. This stems from the fact that the initial value of the output signal Y is unknown and for convenience it is simply assigned the initial value of the input signal X. While the initial value of Y is incorrect, it allows the calculation to proceed. With four passes through the algorithm, the steady-state output is only achieved after about the third cycle of the input

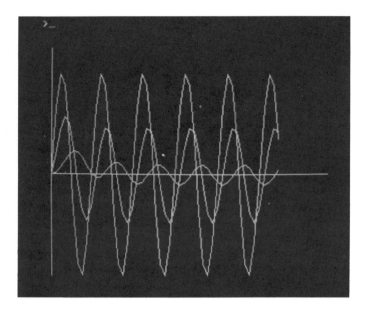

Figure 2.36   *Digital filter output*

signal. The transient state damps out more rapidly with a single pass, with the steady-state reached virtually after the first cycle.

While the program given serves to illustrate the principle of digital filtering, a much more useful exercise is to investigate the performance of a digital filtering algorithm using a real voltage input. This can be easily produced from a waveform generator having control functions on both frequency and amplitude. Using the analogue input terminal available on the BBC microcomputer, see section 7.1, the sampling frequency can be made to approach 270 Hz by selecting channel 1 only and 8-bit resolution. The input signal must be uni-polar and may only vary over the range 0 to 1.8 volts. If a sinusoidal input is to be examined, then it will require rectification to produce a positive only signal. Alternatively an IBM-PC, or compatible machine, with a suitable A/D expansion card, see section 8.4, may also be used to observe the effect of the digital filter.

An illuminating study can then be made on the effects of varying input frequencies from say, 5 Hz to 30 Hz, with a selection of filter cut-off frequencies, say 1 Hz, 10 Hz and 20 Hz, and a varying number of passes through the algorithm.

The highpass RC filter algorithm can be studied in much the same manner. The discretised version of equation (2.11) takes the form:

$$Y_i + \frac{\tau}{\Delta t} (Y_i - Y_{i-1}) = \frac{\tau}{\Delta t} (X_i - X_{i-1})$$

$$\therefore Y_i = \frac{1}{\left(1 + \frac{\Delta t}{\tau}\right)} [Y_{i-1} + X_i - X_{i-1}] \tag{2.17}$$

While electronic noise may be the most significant problem associated with low-level signals, impedance mis-match is also frequently encountered. The impedance mis-match problem can be illustrated by considering a voltage output transducer with an internal impedance, figure 2.37.

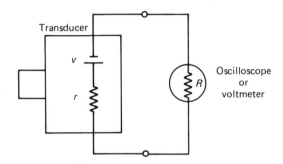

Figure 2.37 *Impedance mis-match problem*

For any particular input conditions, the transducer functions as a small voltage source, v, in series with a resistance r. The voltage is measured with a voltmeter having an internal resistance, R.

The current flowing along the circuit is $i = \dfrac{v}{(r+R)}$

and the voltage recorded by the meter is therefore $v - ir$, that is

$$v - \frac{vr}{(r+R)} = v\left(\frac{r+R-r}{r+R}\right) = \frac{vR}{(r+R)}$$

$$= \frac{v}{(1+r/R)} \qquad (2.18)$$

The error in the measurement is ir

$$= \frac{vr}{(r+R)}$$

$$= \frac{v}{(1+R/r)} \qquad (2.19)$$

This is simply another manifestation of the loading error which was discussed in section 2.2. If $R \gg r$, then the error may be inconsequential. The sensitivity of the measurement however may be considerably reduced if $r \gg R$.

The problem may be greatly alleviated by interfacing a high input impedance, low output impedance amplifier between the transducer and the recording equipment. Amplifiers of this sort are referred to as 'buffer' amplifiers, or just simply 'buffers'. Figure 2.38 depicts the basic layout in schematic form.

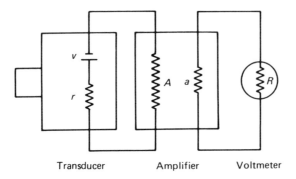

Figure 2.38 Buffer amplifier

For particular values R, A, r and a of 10 kΩ, 200 MΩ, 500 kΩ and 200 Ω respectively, equation (2.19) gives the error associated with the first stage as approximately 0.25%. The second stage results in an error of about 2%.

With the figures as given, the cumulative errors seem to be unacceptably high. They should be viewed however in comparison with the situation which would exist in the absence of the buffer amplifier, where the attenuation of the signal would approach 100%. The gain of the amplifier has not been taken into account in the example given, but this is usually unity: the primary function being the prevention of serious signal loss. Figure 2.39 shows the general purpose '741' being used as a non-inverting, unity-gain buffer amplifier.

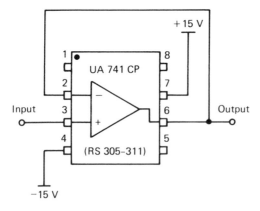

Figure 2.39 Unity-gain buffer amplifier

2.6 CALIBRATION, CURVE FITTING

In many of the applications considered in this chapter, reference has been made to the requirement for calibration. Calibration is the procedure whereby a measured quantity, related to some physical variable, is compared against a known standard. A functional relationship may then be established and subsequently used as a calibration curve. British Standards, BS 4937 for example, give high order polynomials to represent thermoelectric voltages for temperatures ranging between $-270°C$ to $0°C$ and $0°C$ to $1400°C$. For many other applications there are no such convenient standards available.

Approximating polynomials may however be derived, through various numerical procedures, to fit an appropriate curve through the experimental data. A popular technique involves the so-called 'least-squares' averaging method to fit the 'best' approximating polynomial through measured data. The method, which has great utility in the analysis and correlation of experimental data, is described in some detail in appendix II along with the listing of a general program in BASICA to fit an arbitrary polynomial of up to order 10 through any set of measured points.

EXERCISES

1. A variable potential divider has a total resistance of 2 kΩ and is connected to a 10 volt d.c. supply as shown.

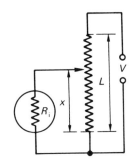

The internal resistance of the voltmeter is 5 kΩ.
Determine the loading error for $x/L = 1/4, 1/2, 3/4$.
Note: the loading error is the difference between that which would be measured if the voltmeter had infinite internal resistance and the actual measured voltage with $R_i = 5$ kΩ.

[0.174 V, 0.455 V, 0.523 V]

2. A strain bridge comprises two fixed 130 Ω resistors, one active gauge and one unstrained gauge for temperature compensation. The two gauges have un-

strained resistances of 130 Ω and gauge factor of 2.0. Determine the bridge output for a bridge supply of 5 V when the active gauge is subject to 800 microstrain.

[2.0 mV]

3. A strain bridge is made up of two fixed 120 Ω resistors and two active gauges subject to equal and opposite strains on opposite surfaces of the flexural member. The two gauges have unstrained resistances of 120 Ω and gauge factor of 2.5. The bridge is supplied with 6 V and when strained, the bridge output is 1.6 mV. Determine the microstrain developed in the gauges.

[213.33 micro-strain]

4. Strain gauges are to be mounted on to a loaded cantilever to function as a simple force transducer. Show that by using active gauges on all four arms of the bridge circuit, the output voltage may be given by the relation:

$$v = V(\Delta R/R)$$

where V is the bridge supply voltage
         ΔR is the change in gauge resistance caused by the load
         R is the unstrained gauge resistance.

Indicate schematically how the gauges would physically be wired up and attached to the cantilever.

5. A simple RC circuit is to be used as a high pass filter with a 3 db cut-off frequency of 10 Hz. Determine the required time constant and specify suitable sizes for the capacitor and resistor.

[0.0159 s, 1.0 μF, 15.9 kΩ]

6. A bandpass filter is to be constructed using the circuit diagram shown below:

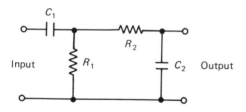

If the unity gain bandwidth is to extend from 200 Hz to 10 kHz, determine the required time constants and specify suitable values for the circuit components R_1, C_1, R_2 and C_2.

[7.96 × 10^{-4} s, 1.59 × 10^{-5} s; 796 Ω, 10 μF, 0.1 μF, 159 Ω]

7. Derive the finite difference approximation for a simple RC high pass filter, i.e. equation (2.17).

The analogue output from a transducer is sampled and converted to digital form at the rate of 20 kHz. The frequency range of interest is 0 to 4 kHz and it is therefore intended to filter the digitised signal with a 3 db cut-off frequency at 5 kHz. Determine the required value of the constant A in the digital filtering algorithm:

$$Y_i = A \left[Y_{i-1} + X_i - X_{i-1} \right]$$

[0.389]

8. A particular transducer has an internal resistance of 1.0 kΩ. The transducer output voltage is recorded on a digital voltmeter having an internal resistance of 100 MΩ. Determine the percentage error in the voltmeter reading.

[0.0010%]

REFERENCES

Barney, G. C., *Intelligent Instrumentation*, 1st edn, Prentice-Hall, 1985.

Bass, H. G., *Introduction to Engineering Measurements*, McGraw-Hill, 1971.

British Standards Institution, *BS 1041, Temperature measurement*, various dates.

British Standards Institution, *International Thermocouple Reference Tables, BS 4937, part 5: 1974* and *part 20: 1983*.

Doebelin, E. O., *Measurement Systems, Application and Design*, 2nd edn, McGraw-Hill, 1983.

Hayward, A. T. J., *Flowmeters —a Basic Guide and Source Book for Users*, Macmillan, 1983.

Holman, J., *Experimental Methods for Engineers*, McGraw-Hill, 1984.

Measurement Group Educational Div., *Student Manual for Strain Gauge Technology*, Welwyn Strain Measurement, Armstrong Rd, Basingstoke, England.

National Physical Laboratory, *The International Practical Temperature Scale of 1968*, 1976.

Oliver, F. J., *Practical Instrumentation Transducers*, Pitman, 1972.

Williams, A. B., *Electronic Filter Design Handbook*, McGraw-Hill, 1981.

Chapter 3
Error Accountability

3.1 ERRORS IN EXPERIMENTAL MEASUREMENTS

The degree of uncertainty in any dependent variable can only be found by an examination of all the sources and contributory errors associated with the independent variables upon which the dependent is subject. It is not always possible to make an exhaustive study of every measurement taken and so a global assessment, based on the evidence available, is normally stated. The end result should be a conclusion to the effect that the value indicated by the measurements taken is unlikely to have differed from the true value by more than some stipulated amount. Both systematic and random errors can accrue in experimental measurements and each must be taken fully into account.

Random errors are indicated by the scatter in repeated measurements of the same variable about the mean value. If the error is truly random in the strictest sense, then the frequency distribution of the deviation between the measured variables and the mean value will be Gaussian. Systematic errors, on the other hand, are related to the inherent accuracy of the instrumentation used and the limitations of their calibration. To a reasonable extent, systematic errors can be accounted for through accurate calibration. For example, if a pressure gauge is known to read high in a consistent manner then the systematic error can be eradicated by application of the appropriate correction factor to each individual measurement. Random errors must be dealt with in a different manner.

3.2 UNCERTAINTY ESTIMATING

If a set of n repeated observations of a quantity have a random distribution about the mean value, then the arithmetic mean of the observations is given by:

$$M = \frac{O_1 + O_2 + O_3 + \ldots + O_n}{n} \tag{3.1}$$

where O_n are the numerical values of the observed measurements.

The standard deviation of the observations is the root mean square value of the deviations about the mean and is a measure of the spread of the variation. Mathematically, the standard deviation is written as:

$$\sigma = \pm \left[\frac{(O_1 - M)^2 + (O_2 - M)^2 + (O_3 - M)^2 + \ldots + (O_n - M)^2}{n} \right]^{1/2} \quad (3.2)$$

As n increases, the value of M tends to approach the true mean value. Provided n is sufficiently large (greater than about 30) and the variations are normally distributed, it would be found in general that 95 per cent of the readings would be grouped about the mean value within limits which are related to the standard deviation, approximately 1.96σ. Clearly, 99 per cent of the readings would be contained within wider limits, similarly related to σ, but for most practical purposes the 95 per cent band is adopted as reference. This is referred to as the 95 per cent confidence limits and the intelligent bookmaker might be prepared to lay odds of 19 to 1 that any single measurement would lie within these limits. If say 100 gambles were taken with stakes of £1.00, then the bookmaker would perhaps win 95 times. It is possible that the bookmaker would lose 5 times and have to pay out £19.00 each time and thereby break even. By reducing the odds to 18 to 1, the bookmaker can be assured of a net profit.

This concept of a bookmaker's odds is particularly useful as a means of describing the uncertainty in experimental measurements. While the actual uncertainty may not necessarily be normally distributed, careful consideration of all the factors contributing to error can allow the experimenter to lay odds that a particular measurement will lie within a prescribed range. Representing the range by w and the odds by b, the uncertainty becomes:

$$M \pm w \ (b \text{ to } 1) \quad (3.3)$$

As a practical example, one might write

Pressure $= 100.3 \pm 1.0 \text{ kN/m}^2$ (20 to 1)

The above states that the best estimate pressure is 100.3 kN/m^2 and that the odds are 20 to 1 that the true value lies within $\pm 1.0 \text{ kN/m}^2$ of this measurement. The range w is selected such that the experimenter would be willing to wager b to 1 that the error would be less than w. Determination of w may in itself be based on an estimate, but it is the responsibility of the experimenter to make the best effort possible; no-one else is in a better position to do so.

3.3 DIGITAL UNCERTAINTY

With the arrival of computer technology to experimentation a new source of error has manifested in the form of digital uncertainty. This is associated with the conversion of analogue signals to digital representation and is not an error in the strictest sense, but more properly termed the resolution. The digital uncertainty will depend on the type of A/D converter used, see section 6.3. If the reader is unfamiliar with the binary numbering system, then reference should be made to section 4.1 before proceeding here.

The concept of digital uncertainty is best illustrated by a practical example for which we can consider the conversion of an analogue signal by an 8-bit A/D converter, having a reference voltage of 2.55 V. It can be assumed that the analogue signal represents some physical variable and has been conditioned to range between 0 and 2.55 V. The full numerical range of the 8-bit converter is 0 to 255 and if at some stage the analogue voltage is say 1.827 V, then the conversion would be approximated as shown below:

Numerical value corresponding to input voltage = $(1.827/2.55)\ 255 = 182.7$

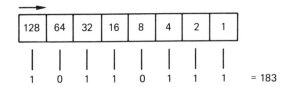

The digitised voltage is therefore equivalent to 1.83 V.

The last digit in the numerical value may be 3 or 2, depending on whether the converter rounds up or down respectively. It can be seen therefore that the least significant bit in the digital representation is uncertain and that the absolute accuracy of the digital approximation is 1 in 255 of the full scale reading. Normally manufacturers would quote a resolution to ± half of the least significant bit as a result of the round-off error. By considering the absolute accuracy to be ± the least significant bit, the estimated error is more conservative and can allow for other error sources including non-linearity or zero drift which are both associated with the electronic circuitry.

Since the signal represents a physical variable, then the digital uncertainty can easily be related back to the consequential uncertainty in the actual variable. For example, if the signal emanates from a displacement transducer, with displacement in the range 0 to 250 mm, then the absolute resolution of the displacement is to $\pm(1/255) \times 250$, i.e. approximately ±1 mm.

To improve on this resolution a 12-bit A/D converter may be used with a numerical range of 0 to 4095. The reference voltage of the 12-bit converter may well be the same as the previous 8-bit converter and with the same analogue input, the numerical value corresponding to the input voltage is:

$(1.827/2.55) \times 4095 = 2933.95$

The approximation is as illustrated below:

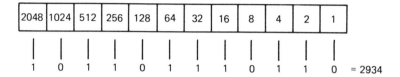

The digitised voltage in this case is represented as $(2934/4095) \times 2.55$
$$= 1.8270.$$

It can be seen that the resolution has been improved by a factor of about 10, with the digitised voltage now accurate to about ±0.001 V.

Obviously with a 16-bit A/D converter the resolution can be further improved but this will be at the expense of perhaps conversion time and most certainly cost, both of which may be limiting factors.

Consideration of the resolution allows a statement of the resulting uncertainty in the measured physical parameter to be made in much the same manner as for any other error source, i.e. equation (3.3). The resolution is stated as ±1 bit in m of the full scale reading, where m is the maximum decimal number associated with the number of bits of the converter (see section 6.3 for fuller details).

Example

In an experimental investigation of a vibrating cantilever, a linear displacement transducer is used to measure the amplitude of oscillation of the free end of the cantilever. On amplifying the transducer output, a linear relationship is obtained between voltage and displacement where 8.90 volts corresponds to a maximum physical displacement of 6.5 mm.

An 8-bit A/D converter is used to digitise the signal and the internal reference voltage of the converter is 10.0 volts.

Determine the resolution of the digitised voltage and the resulting uncertainty in the measured displacement.

Resolution of the digitised voltage = $(1/255) \times 10.0 = \pm 0.03922$ volts.

Full scale reading = $(10.0/8.9) \times 6.5 = 7.3034$ mm.

Uncertainty in displacement = $(1/255) \times 7.3034 = \pm 0.02864$ mm.

The accuracy of the displacement is therefore = $(0.02864/6.5) \times 100\%$
$$= \pm 0.4406\%.$$

If a 12-bit A/D converter was to be used, having the same reference voltage, then:

Resolution of the digitised voltage = $(1/4095) \times 10.0 = \pm 0.002442$ volts.

Uncertainty in displacement = $(1/4095) \times 7.3034 = \pm 0.001783$ mm.

The improved accuracy in the measured displacement = $(0.001783/6.5) \times 100$
$$= \pm 0.0274\%.$$

3.4 ERROR ANALYSIS

Equation (3.3) gives a standardised method of describing uncertainties in each of the basic variables. The next requirement is to determine how these uncertainties may propagate into derived quantities. To illustrate the error analysis techniques, they will be applied to the practical example of the measurement of air velocity by a Pitot-static tube. The theory and operating principle of the Pitot-static tube is covered in most basic texts on Fluid Mechanics and its use is standardised in British Standard 1042. The governing equation for incompressible flow is:

$$\text{velocity, } C = 6.506 \sqrt{\left(\frac{h_1 T}{h_2}\right)} \tag{3.4}$$

where the physical measurements are:

h_1 — the level difference, in mm, on a U-tube manometer containing water and connected to the stagnation and the static pressure tappings of the Pitot-static tube.

h_2 — the level difference, in mm, on an atmospheric barometer containing mercury.

T — the ambient temperature in degrees Kelvin.

The constant appearing in the relation is related to other physical constants, including the characteristic gas constant for air (287 J/kg K), the density of fresh water (10^3 kg/m^3) and the density of mercury (135,600 kg/m^3).

The result is the fluid velocity, in m/s, and although the three basic measurements are simple, there are a great many sources of possible error. These include the alignment of the device to the oncoming flow, leakage in the pressure tubing, non-uniformities in the U-tube bore, fluctuations in atmospheric conditions, unaccounted compressibility effects and the accuracy in reading of the three variables themselves.

For compatibility, the range for the three variables must be based on the same odds and this is usually 20 to 1 since it is closely associated with the 95 per cent confidence limits. If T is measured with a mercury-in-glass thermometer, h_1 with an inclined water manometer and h_2 with a Fortin barometer, then a description of the uncertainty might be:

$h_1 = 30 \pm 0.2$ mm H$_2$O (20 to 1)
$T\ \ = 293 \pm 0.5$ degrees K (20 to 1)
$h_2 = 758 \pm 0.5$ mm Hg (20 to 1)

(i) The Method of Kline and McClintock

In this method, the dependent variable, P, is assumed to be a linear function of n independent variables, $v_1, v_2, \ldots, v_n$, such that:

$$P = P(v_1, v_2, \ldots, v_n) \tag{3.5}$$

For small variations in the independent variables, the total variation in the dependent variable is:

$$\delta P = \frac{\partial P}{\partial v_1} \; \delta v_1 + \frac{\partial P}{\partial v_2} \; \delta v_2 + \ldots + \frac{\partial P}{\partial v_n} \; \delta v_n \qquad (3.6)$$

Provided that the variations in the independent variables are normally distributed, then the range for the dependent variable can be related to that for each of the independent variables, with the same odds, by:

$$\omega_p = \left[\left(\frac{\partial P}{\partial v_1} \; \omega_{v_1} \right)^2 + \left(\frac{\partial P}{\partial v_2} \; \omega_{v_2} \right)^2 + \ldots + \left(\frac{\partial P}{\partial v_n} \; \omega_{v_n} \right)^2 \right]^{1/2} \qquad (3.7)$$

In applying this method to the air velocity measurement problem, it is convenient to re-express equation (3.4) in the logarithmic form as follows:

$$\ln C = \ln A + \tfrac{1}{2} \ln \left(\frac{h_1 T}{h_2} \right) \qquad (3.8)$$

or $\qquad \ln C = \ln A + \tfrac{1}{2} \ln h_1 + \tfrac{1}{2} \ln T - \tfrac{1}{2} \ln h_2 \qquad (3.9)$

where $A = 6.506$.
Differentiation gives:

$$\left(\frac{1}{C} \frac{\partial C}{\partial h_1} = \frac{1}{2h_1} \right) ; \; \left(\frac{1}{C} \frac{\partial C}{\partial T} = \frac{1}{2T} \right); \; \left(\frac{1}{C} \frac{\partial C}{\partial h_2} = - \frac{1}{2h_2} \right)$$

and substitution into (3.7) results in:

$$\frac{\omega_c}{C} = \left[\left(\frac{1}{2} \frac{\omega_{h_1}}{h_1} \right)^2 + \left(\frac{1}{2} \frac{\omega_T}{T} \right)^2 - \left(\frac{1}{2} \frac{\omega_{h_2}}{h_2} \right)^2 \right]^{1/2} \qquad (3.10)$$

Note that the negative sign qualifying the last group of terms has no particular significance since the range, w, can be either positive or negative.

Hence $\dfrac{\omega_c}{C} = \dfrac{1}{2} \left[\left(\dfrac{0.2}{30} \right)^2 + \left(\dfrac{0.5}{293} \right)^2 + \left(\dfrac{0.5}{758} \right)^2 \right]^{1/2}$

$$= \tfrac{1}{2} \, [4.44 \times 10^{-5} + 2.91 \times 10^{-6} + 4.35 \times 10^{-7}]^{1/2}$$

$$= 0.00346$$

or $\qquad \dfrac{\omega_c}{C} = \pm \, 0.346\%$

The analysis shows that the velocity C can be expected to have an accuracy of ±0.35% of the true value. Using the given data this results in a range of C from 22.078 m/s to 22.232 m/s with the mean value being 22.155 m/s. The analysis also indicates where the largest contribution to the error occurs. In the above case it is the measurement of h_1 which constitutes the major source of error. It should be apparent therefore that there would be little point in trying to improve the accuracy of measurement of the other two variables, T and h_2, until the accuracy of measurement of h_1 is considerably improved.

(ii) Binomial Expansion

Written generally, the Binomial expansion is:

$$(a + b)^n = a^n + na^{n-1}b + \frac{n(n-1)a^{n-2}b^2}{2!} + \frac{n(n-1)(n-2)a^{n-3}b^3}{3!} +...$$

and for the special case where $a = 1$ and $b = x$, the expansion becomes: (3.11)

$$(1 + x)^n = 1 + nx + \frac{n(n-1)x^2}{2!} + \frac{n(n-1)(n-2)x^3}{3!} + ...   \qquad (3.12)$$

The proof of equations (3.11) and (3.12) are given in many standard mathematical texts, to which the reader is referred.

   In applying the theorem to an error analysis of the air velocity measurement, it is convenient to express the error in each measurement as follows:

$$h_1' = h_1 \left(1 + \frac{0.2}{30}\right) = h_1 (1 + 0.00667)$$

$$T = T \left(1 + \frac{0.5}{293}\right) = T (1 + 0.00171)$$

$$h_2' = h_2 \left(1 + \frac{0.5}{758}\right) = h_2 (1 + 0.00066)$$

where the prime (') denotes the measured value in terms of its deviation from the true value.

Thus $C' = 6.506 \sqrt{\left(\dfrac{h_1' \, T'}{h_2'}\right)}$

$$= 6.506 \left[\frac{h_1 (1 + 0.00667) \, T (1 + 0.00171)}{h_2 (1 + 0.00066)}\right]^{1/2}$$

$$= C \left[\frac{(1 + 0.00667)(1 + 0.00171)}{(1 + 0.00066)}\right]^{1/2}$$

$$= C [(1 + 0.00667)^{1/2} (1 + 0.00171)^{1/2} (1 + 0.00066)^{-1/2}$$

   Since the higher order terms of x are increasingly negligible, it is normally sufficient to include only the first two terms in each expansion.

Hence $C' = C[(1 + \frac{1}{2} \times 0.00667) (1+ \frac{1}{2} \times 0.00171) (1 - \frac{1}{2} \times 0.00066)]$

$$= C[(1 + 0.00333) (1 + 0.00086) (1 + 0.00033)]$$

Note again, that since each error can be either positive or negative, then there is no particular significance attached to the minus sign inside the last bracketed term ( ).

Multiplying the expansion out and neglecting the product of small order terms results in:

$$C' = C \, [1 + 0.00452]$$

The above shows that the velocity C has an expected accuracy to $\pm$ 0.452% and is similar to the result that was obtained with the previous analysis.

In a similar manner, the expansion method also clearly identifies the most significant source of error as in the measurement of h_1.

(iii)   Monte Carlo Method

The Monte Carlo technique utilises the function of random number generation which is a feature common to most microcomputers. In the method the independent variables are generated randomly, but all within the prescribed error range. The dependent variable is then calculated from the governing equation and the normal statistical analyses performed on these calculated results. Using say, 1000 randomly generated results, the arithmetic mean, standard deviation and 95 per cent confidence limits are all easily obtained.

The advantages of this method over the more conventional techniques are:
1. No assumptions are made regarding the functional relationship between the dependent and independent variables.
2. No assumptions of the negligibility of second or higher order terms are required.
3. The inter-relationship between variables are automatically accounted for.
4. It is easy to alter the range on any measured variable to ascertain the effect on the resulting error.
5. There is no limitation on the complexity of the relationship between the dependent and independent variables.

The technique can best be illustrated through its application to the air velocity measurement problem previously considered. The program listing is given below:

```
10 REM ERROR ANALYSIS BY A MONTE CARLO METHOD
20
30 DIM C(1000)
40 REM RANDOM GENERATION OF 1000 RESULTS
45 CLS
50 PRINT"WAIT, COMPUTATION IN PROGRESS":PRINT
60 FOR K=1 TO 1000
70 H1=29.8+RND(40)/100
80 H2=757.5+RND(100)/100
90 T=292.5+RND(100)/100
100 C(K)=6.506*SQR(H1*T/H2)
110 NEXT K
```

```
120 PRINT"RANDOM RESULTS GENERATED":PRINT
130 REM STATISTICAL ANALYSIS
140 SUM=0
150 ST=0
160 FOR K=1 TO 1000
170 SUM=SUM+C(K)
180 NEXT K
190 MEAN=SUM/1000
200 FOR K=1 TO 1000
210 ST=ST+(C(K)-MEAN)*(C(K)-MEAN)
220 NEXT K
230 STD=SQR(ST/1000)
240 PRINT:PRINT"Arithmetic Mean = ";MEAN
250 PRINT:PRINT"Standard Deviation = ";STD
260 CL95=1.96*STD
270 PRINT:PRINT"95 % Confidence Limits = ";CL95
280 LOW=MEAN-CL95
290 HIGH=MEAN+CL95
300 KOUNT=0
310 FOR K=1 TO 1000
320 IF C(K)<LOW GOTO 350
330 IF C(K)>HIGH GOTO 350
340 KOUNT=KOUNT+1
350 NEXT K
360 PRINT:PRINT"Percentage of Results within Confidence Limits = ";KOUNT/10
370 A=196*STD/MEAN
380 PRINT:PRINT"Accuracy = ";A;" %"
```

In running the program a typical output is as follows:

WAIT, COMPUTATION IN PROGRESS
RANDOM RESULTS GENERATED
Arithmetic Mean = 22.1591742
Standard Deviation = $4.3878397E - 2$
95% Confidence Limits = $8.60016581E - 2$
Percentage of Results within Confidence Limits = 98.9
Accuracy = 0.388%

Execution of the program on a microcomputer takes about 57 seconds with the bulk of the time utilised in the random generation of the 1000 samples. Slightly different results are obtained with every re-run of the program, but the variation is well within acceptable limits and the final result is of the same order of magnitude as that obtained with the other two methods.

The program is in fact quite general and only requires lines 70 to 100 to be altered to suit the user's requirements. In this section of the program, each variable is set to the lowest value within the range and the random number generated to cover both the negative and the positive parts of the range. The only proviso is that the computed array variable is always denoted by C(K).

It is difficult to make a recommendation as to which method should be preferred since they all have their points for and against. The conventional methods, (i) and (ii), both give a clear indication of the dominant error in the analysis, but the Monte Carlo method is attractive in its ease and simplicity. For this reason we tend to favour the numerical Monte Carlo method. If nothing else it is the most modern technique and places a minimal demand on mathematical skills.

3.5   REPRESENTATION OF UNCERTAINTY

Apart from a bland statement of the accuracy of any result a more useful indication can often be presented in graphical form. Figure 3.1 illustrates one particular means of depicting the uncertainty band pertaining to measured data.

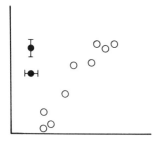

Figure 3.1

Figure 3.1 shows the relationship between two variables, both of which are derived from primary measurements. Both variables are therefore subject to propagated error and using the methods of section 3.3, the 95 per cent confidence limits can be estimated and shown on the figure to the same scale. For a more dramatic visual indication, the following procedure could be adopted.

1. Draw the best mean curve through the data, figure 3.2a.
2. Superimpose on each side of the mean curve, the uncertainty band for either the horizontal or the vertical axis, figure 3.2b.
3. Superimpose on each extremity of the first band, the uncertainty band for the other axis, figure 3.2c.

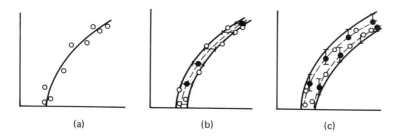

(a)   (b)   (c)

Figure 3.2

On drawing through the outermost extremities, an absolute error band is depicted. If the uncertainty estimates for the two variables have been conservative, then all of the measured data should lie within this absolute error band, figure 3.3.

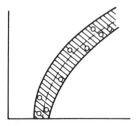

Figure 3.3

EXERCISES

1. The efficiency of a centrifugal pump is given by the relation:

$$\eta = \frac{10^4 QH}{NT}$$

where Q is the volumetric flow rate in litres/s
  H is the head generated in m
  N is the pump rotational speed in rev/min
  T is the applied torque in Newton metres.

Uncertainty estimates for the measured parameters are:
  $Q = 91.0 \pm 2.0$ litres/s (20 to 1)
  $H = 25.0 \pm 2.0$ metres (20 to 1)
  $N = 850.0 \pm 5.0$ rev/min (20 to 1)
  $T = 326.0 \pm 4.0$ Newton metres (20 to 1).

(a) Determine the expected uncertainty in pump efficiency, using:
  (i) the Binomial expansion method
  (ii) the method of Kline and McClintock
  (iii) a Monte Carlo method.

(b) Indicate also, the dominant and the least significant error sources
[±12.0%, ±8.4%, ±9.42%; H dominant, N least significant]

2. An electronic transducer is used as a velocity sensor in the measurement of an air flow.

The calibration for the transducer is given as:

$$C = \left(\frac{V^2 - A}{B}\right)^2 \text{ metres/s}$$

where V is the applied voltage across the transducer sensing element and the constants A and B are 9.331 and 2.753 respectively.
If the estimated uncertainty in the measured voltage is:

$V = 4.50 \pm 0.005$ volts (20 to 1)

determine the resulting uncertainty in the velocity C:
  (i) using the Binomial expansion method
  (ii) using the method of Kline and McClintock
  (iii) using a Monte Carlo method.

[±0.85%, ±0.824%, ±0.92%]

3. The modulus of elasticity of a material, E, can be determined by measuring the deflection of a centrally loaded, simply supported beam, where the governing equation is:

$$E = \frac{WL^3}{4ybd^3}$$

Uncertainty in the measurements are as follows:
load, $W = 100$ Newtons, exact
length, $L = 0.45 \pm 0.001$ metres (20 to 1)
breadth, $b = 0.021 \pm 0.0005$ metres (20 to 1)
depth, $d = 0.005 \pm 0.00005$ metres (20 to 1)
deflection, $y = 0.00413 \pm 0.0001$ metres (20 to 1).
Determine the expected uncertainty in the calculated value for E and indicate the dominant error source:
  (i) using the Binomial expansion method
  (ii) using the method of Kline and McClintock
  (iii) using a Monte Carlo method.

[±8.46%, ±4.58%, ±4.95%; d, dominant]

4. (a) A computerised data-acquisition system is to be used for fundamental research into turbulent flow. The basic physical variables are the mean and the fluctuating flow velocities and these are to be measured by an electronic sensor which has been linearised and conditioned such that 2.0 volts mean corresponds to 20.0 metres/s mean velocity and 2.0 volts r.m.s. corresponds to 0.10 metres/s r.m.s. velocity.
If 8-bit A/D converters, with reference voltages of 2.5 volts are used in the data-acquisition system, determine the uncertainty in the two basic measurements when V_{mean} was 1.6 volts and $V_{r.m.s.}$ was 0.6 volts.
  (b) If the percentage turbulence level is given by the relation:

$T' = 100 \ [C_{r.m.s.}/C_{mean}]\%$

where C refers to the measured velocities, determine the uncertainty in the turbulence level, T':
  (i) using the Binomial expansion method
  (ii) using the method of Kline and McClintock

(iii) using a Monte Carlo method.

$$[16.0 \pm 0.123 \text{ m/s}, 0.03 \pm 0.00049 \text{ m/s}; \pm 2.4\%, \pm 1.81\%, \pm 2.02\%]$$

5. The heat transfer rate to a fluid flowing in a circular pipe is given by the relation:

$$h = \frac{\dot{m} \; Cp \; \ln\left(\dfrac{t_w - t_1}{t_w - t_2}\right)}{\pi DL}$$

where h is the surface heat transfer coefficient in kJ/kg K

   D is the pipe internal diameter in m

   L is the length of the pipe in m

   $\dot{m}$ is the mass flow rate of the fluid in kg/s

   C_p is the fluid specific heat capacity, based on a suitable mean temperature, in kJ/kg K

   t_w is the pipe wall temperature in degrees Centigrade

   t_1 is the fluid inlet temperature in degrees Centigrade

   t_2 is the fluid outlet temperature in degrees Centigrade.

A series of experiments are devised to study the time response of h to a step change in mass flow rate, $\dot{m}$, and the following results were obtained which relate to the initial steady-state condition:

   $\dot{m}$ = 1.50 ± 0.005 kg/s (20 to 1)

   C_p = 4.18 ± 0.01 kJ/kg K (20 to 1)

   D = 0.025 ± 0.0001 m (20 to 1)

   L = 1.50 ± 0.001 m (20 to 1).

All three temperatures were measured by thermocouples and following amplification and other conditioning, a linear calibration was obtained as follows:

   $t = 100 \; V$

where V is the conditioned thermocouple voltage.

The steady-state voltages relating to the initial conditions were:

   t_1 = 0.2 volts

   t_2 = 0.45 volts

   t_w = 1.55 volts

To capture the transient response, the three signals are digitised using 8-bit A/D converters, each with a reference voltage of 1.8 volts.

Using the initial conditions as a basis, determine the expected uncertainty in the heat transfer coefficient, h:

 (i) using the method of Kline and McClintock

 (ii) using a Monte Carlo method.

$$[\pm 2.69\%, \pm 2.95\%]$$

Authors' Note: the logarithmic term in the expression eliminates the Binomial expansion method as a means of determining the uncertainty.

6. A series of tests on the heat transfer problem of the previous question gives the following set of steady-state results:

$\dot{m}$	2.5	2.0	1.5	1.0	0.5	kg/s
h	15.8	13.2	10.9	7.3	4.6	kJ/kg K

These results are to be illustrated graphically in the non-dimensional form, Nu against Re.

where $Nu = \dfrac{hD}{k}$ and $Re = \dfrac{4\dot{m}}{\pi D \mu}$

k is the thermal conductivity of the fluid
and μ is the dynamic viscosity of the fluid.
The uncertainty estimates for k and μ are:
  $k = 0.000625 \pm 0.00002$ kW/m K (20 to 1)
  $\mu = 0.00080 \pm 0.00005$ kg/m s (20 to 1)

Determine the uncertainty in Nu and Re and plot the results indicating the absolute uncertainty band.

[±4.284%, ±0.742%]

Use the uncertainty estimates from the previous question where appropriate.

REFERENCES

Bajpai, A. C., Calus, I. M. and Fairley, J. A., *Statistical Methods for Engineers and Scientists*, Wiley, 1979

British Standards Institution, *BS 1042, section 2.1, Method using Pitot static tube*, 1983

Holman, J., *Experimental Methods for Engineers, 4th edn*, McGraw-Hill, 1984

Kline, S. J. and McClintock, F. A., Describing uncertainties in single-sample experiments, *Mech. Eng.*, Vol. 75, 1953

Robertson, J. A. and Crowe, C. T., *Engineering Fluid Mechanics*, 3rd edn, Houghton Mifflin, 1985

Chapter 4
High and Low Level Programming

Students who are already familiar with number systems, Gray and ASCII codes and the BASIC programming language may wish to omit sections 4.1 to 4.5. Section 4.6 covers a short summary of 6502 assembly language which may likewise be omitted if the reader has some prior experience of programming in this language.

4.1 NUMBER SYSTEMS

In the manipulation of data within a computer *there are no margins for error* and a simple two state numbering system is adopted. This is termed the binary system and it is based on an ON/OFF, HIGH/LOW, LOGIC '1'/LOGIC '0' principle. The physical measure of the states is represented as voltage levels. Ideally for the semi-conductor integrated circuits in a microcomputer system, 5 V denotes a logic '1' while 0 V denotes a logic '0'. In practice a tolerance band is adopted with say 2.4 V to 5 V representing '1' and 0 V to 0.8 V representing '0'.

The micro-electronic devices in the system handle the transfer of information in 1s and 0s termed *bits* which is derived from *binary digit*. A group of eight bits is termed a byte. A number of microcomputers are referred to as 8-bit machines since they handle the transfer of data in 8-bit codes. 16-bit and 32-bit machines are also available.

Computers generally operate with three numbering systems – decimal, binary and hexadecimal. In order to communicate with external devices it becomes necessary to be able to translate between each number system. In a binary system, the only possible numbers are 0 and 1 and the base is chosen as 2.

Consider 8-bits:

bit	7	6	5	4	3	2	1	0
	2^7	2^6	2^5	2^4	2^3	2^2	2^1	2^0
	128	64	32	16	8	4	2	1

most significant bit (MSB)   least significant bit (LSB)

Suppose we have the binary number:

$$1 \quad 0 \quad 0 \quad 1 \quad 1 \quad 0 \quad 0 \quad 1$$

The corresponding decimal number is:

$$128 + 0 + 0 + 16 + 8 + 0 + 0 + 1 = 153$$

The binary value for a decimal number of 37 using the above table, is:

$$0 \quad 0 \quad 1 \quad 0 \quad 0 \quad 1 \quad 0 \quad 1 \text{ i.e. } (32 + 4 + 1)$$

It can be seen that the largest binary number represented by 8 bits is 11111111 = 255 in decimal. Or decimal values in the range 0 to $(2^8 - 1)$ only can be accommodated as an 8-bit binary number, i.e. 0 to 255.

For decimal numbers larger than 255, two bytes can be used giving numbers in the range 0 to $(2^{16} - 1) = 0$ to 65535.

The decimal value corresponding to a 16-bit value can easily be obtained using the above 8-bit conversion technique as follows. Suppose we have the 16-bit binary number:

High byte	Low byte
0 0 0 1 1 0 1 0	1 0 1 1 0 1 0 0

This gives a decimal value of:

$$((16 + 8 + 2) \times 256) + (128 + 32 + 16 + 4)$$
$$= \quad 6656 \quad + \quad 180$$
$$= \quad 6836$$

Note that there is a multiplying factor of 256 between corresponding bits of the high and the low byte, i.e. a binary value of 0000 0001 in the high byte corresponds to the decimal value of 256.

Conversion from decimal to 16-bit binary is performed by first dividing the decimal value by 256. For example, 50395 in decimal to binary:

$$50395/256 = 196 \text{ remainder } 219$$

The 196 represents the decimal value associated with the high order byte and the 219 represents that associated with the low byte.

Thus the 16-bit binary representation is 11000100 11011011.

The direct manipulation of numbers in pure binary form is extremely cumbersome and as you can imagine, 32-bit numbers written in this form are unthinkable. For this reason a shorthand numbering system for binary is adopted which forms the number as groups of 4 bits. 4 bits, which make up half a byte, is termed a nibble and the byte is split into an upper and lower nibble. The 4 bits are then labelled as one alpha-numeric value which will represent decimal numbers in the range 0 to $(2^4 - 1) = 0$ to 15. This numbering system, formed to a base of 16, is termed hexadecimal, or simply hex for short. Since there are only ten unique symbols used in our decimal numbering system, the first six letters of the alphabet are used to denote the consecutive decimal numbers from 10 to 15, that is:

Decimal	0	1	2	3	4	5	6	7	8	9	10	11	12	13	14	15
Hexadecimal	0	1	2	3	4	5	6	7	8	9	A	B	C	D	E	F

Thus 8-bit binary numbers may be replaced by 2 hex digits. For example:

14 decimal = 0000 1110 binary = 0E hex

Consider the decimal number 156. To split the number into its upper and lower nibbles we can divide by 16, that is:

156/16 = 9 remainder 12

The 9 is the decimal value associated with the upper nibble, while the 12 is that associated with the lower nibble.
Thus 156 decimal = 1001 1100 binary = 9C hex.

For a decimal number which requires 16 bits to be represented in binary form we can first of all divide the number by 256. This will split the number into the decimal values associated with the upper and lower bytes. Dividing these numbers each by 16 will then split the bytes into their respective upper and lower nibbles. For example:

Consider the decimal number 6836
6836/256 = 26 remainder 180
26/16 = 1 remainder 10 and 180/16 = 11 remainder 4

Thus the decimal numbers associated with the four nibbles which make up the 16-bit number are:

```
                    1    10   11   4
Thus 6836 decimal = 0001 1010 1011 0100 binary
                  =   1    A    B    4   hex
```

In a similar manner, decimal numbers greater than 65535 can be handled in 32-bit binary notation.

The convenience of the shorter form hexadecimal numbering system becomes apparent and various microcomputers can handle both decimal and hex values for input or output of data. The default, or assumed representation, is usually

decimal and hex numbers are preceded by an H, &H, $ or &. The BBC micro-
computer uses the latter and the IBM-PC uses &H. Conversion between decimal
and hexadecimal, or vice-versa, can be easily carried out at the keyboard.

For example, for the BBC microcomputer:

type: PRINT ~6836 ⟨CR⟩
where '~' is a tilde, obtained by SHIFT ^

This returns the hexadecimal value of a decimal number.
Also try:

PRINT &1AB4 ⟨CR⟩

The equivalent commands in IBM's BASICA are:

PRINT HEX$(6836)
and   PRINT &H1AB4

The hex numbering system is by far the most important in microcomputers.
It can easily be translated into binary, from which the corresponding decimal
number can be evaluated as shown previously.

The addition of binary numbers can be summarised in the following simple
rules:

0 plus 0 gives 0
0 plus 1 gives 1
1 plus 0 gives 1
1 plus 1 gives $\overline{0}$

In the last addition the result is $\overline{0}$ and the overscore denotes that a 1 must be
carried over and added to the numbers in the next most significant bits. For
example:

13 + 7 = 20 decimal

Performing the addition in binary notation, we have:

13	1101
+7	+0111
= 20	= 10100

In adding the numbers in the right-hand column, the numerical result is 0 and a
1 must be carried to the next column. Adding the numbers in the second column
we have 0 plus 1, which is 1 and this 1 must then be added to the 1 which was
carried over from the previous column. The final result is again 0 and a 1 is again
carried over to the third column. For the third column, the first addition will
result in 0 with a carry of 1 to the fourth column. The second addition of 0 and
1 will result in a 1 being assigned to the third column of the answer. The addition
of the fourth column results in 0 with a carry of 1 to the fifth column.

Negative numbers must obviously be handled by the system and this is performed by using what is termed the 'two's complement' technique. The most significant bit of the binary number is used to represent the sign. A '1' denotes a negative number and '0' a positive number. Since the most significant bit is used to indicate the sign, then the numerical value can only be represented with the remaining 7 bits. This restricts the range of decimal values which can be handled from −128 to +127. The corresponding decimal range for a signed 16-bit binary number is −32768 to +32767. For example, 10101100 will be a negative signed binary number.

The technique operates as follows:

to represent say −104 decimal as a signed binary or hex value
decimal value, +104 = 0110 1000 binary

the complement   = 1001 0111 binary
and add 1   0000 0001
                     ─────────
gives   1001 1000

It is noticed that the MSB is a 1 thus denoting a negative number. Hence −104 = 1001 1000 in signed binary.

    =   9   8   in signed hex.

This can be checked by adding +104 to −104 in binary notation.

+104   0110 1000
−104   1001 1000
        ─────────

add   0   10000 0000

The result is zero in 8-bit representation. Note that although a carry bit is displayed it will not be registered since only the first 8 bits are recognised.

The conversion from signed binary to decimal is similar to the above. For example:

            1110 0010 is a negative number in signed binary notation

complement 0001 1101
add 1   0000 0001
            ─────────
            0001 1110

Since 0001 1110 in binary represents 30 in decimal, then 1110 0010 in signed binary represents −30 in decimal.

As a further illustration, consider the decimal subtraction 37 − 14. The answer is obviously 23 but it is informative to consider how this answer can be arrived at in binary notation. The easiest method is to convert the 14 to −14 in signed binary, using the 2's complement and then to perform the binary addition of 37 + (−14).

Thus 14   = 0000 1110

the complement = 1111 0001
and add 1   0000 0001
gives (−14)   = 1111 0010

now 37   = 0010 0101
+ (−14)   1111 0010
             1 0001 0111

Since only the first 8-bits are read, then the answer is correct.

In many microcomputer systems, integer quantities are often handled as four bytes with the most significant bit indicating the sign of the number. This leaves 31 bits to represent the decimal number and results in integers in the range of -2^{31} to $(2^{31} - 1)$. The corresponding range in decimal is:

$$-2\ 147\ 483\ 648 \text{ to } +2\ 147\ 483\ 647$$

Real numbers are stored as 5 bytes with the most significant byte being the exponent and a sign bit, and the other 4 bytes being the mantissa and a sign bit.

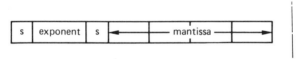

Considering the most significant byte, a '1' in the most significant bit denotes a negative exponent. In 8-bit signed binary notation the decimal number for the exponent may then range only between −128 and +127. This confines the range of the exponent to:

$$2^{-128} \text{ to } 2^{127}$$

The mantissa gives the floating point decimal value and again the most significant bit is used to denote the sign. The mantissa is formed as shown below:

MSB						LSB
s	2^{-1}	2^{-2}	2^{-3}	2^{-4}		2^{-31}

The manipulation of positive and negative floating point numbers within the microcomputer system is quite complex. For most practical purposes it is sufficient to be aware that the range of real numbers which can be handled by the system is:

$$\pm 0.15 \times 10^{-38} \text{ to } \pm 1.7 \times 10^{38}$$

In addition, since the numerical value of the mantissa is represented in 31 bits, then the accuracy of any real number is better than 9 significant digits, but less than 10.

Note: electronic calculators use a different numbering system and are capable of handling numbers in the larger range of 10^{-99} to 10^{99} with a specified number of digits.

4.2 GRAY CODE

This is one of many binary type codes in which only one of the digits change between successive numbers. For this reason it is often referred to as 'a unit distance code' and is mainly applicable in expressing the mechanical position of a mechanism in binary terms. Either rotational or translational position can be measured by means of an encoder which utilises a Gray code. Details of an absolute shaft encoder are given in section 2.1. The Gray code compared with decimal and binary is shown in table 4.1.

Table 4.1   Binary and Gray codes

Decimal	Binary	Gray
0	0000	0000
1	0001	0001
2	0010	0011
3	0011	0010
4	0100	0110
5	0101	0111
6	0110	0101
7	0111	0100
8	1000	1100
9	1001	1101

Various methods are available for detecting the position from the encoder but the optical method of reading is the most popular. Beams of light are focussed onto the tracks which are alternatively opaque and transparent, and the light output on the other side of the encoder is detected by photocells to produce a digital output signal. It is seen from the pure binary code that on 4 bits, all of the digits change between the decimal values of 7 and 8. Any misalignment of optical sensors could therefore result in an incorrect code denoting a possible 180 degree error in actual rotational position. The unit-distance Gray code eliminates this possible 180 degree error and limits it to a 1 bit equivalence in degrees.

The number of tracks, or bits, on the encoder defines the resolution achievable. For example, in order to measure to at least 1^0 of movement a 1 in 360 resolution is required. This necessitates at least a 9-bit encoder. (2^9 = 512, giving a measurement accuracy of 0.7^0.)

Although a Gray code is used on an encoder, the computer operates in binary and a code conversion algorithm for Gray to binary is necessary when inputting the encoder output data to a microcomputer.

A suitable algorithm operates as follows;

The most significant bit (MSB) of the binary number B, equals the MSB of the Gray code G. Thereafter

$$Bn = Gn \oplus B(n + 1)$$

where n denotes the bit number and $\oplus$ stands for an exclusive-OR logic command operating on two logic values. The truth table for an exclusive-OR (EOR) is:

$0 \oplus 0 = 0$
$0 \oplus 1 = 1$
$1 \oplus 0 = 1$
$1 \oplus 1 = 0$

For example, to convert 1101 in Gray to binary:

```
bit     3  2  1  0
Gray    1  1  0  1
        |  /  /  /
         ⊕  ⊕  ⊕
        ↓  ↓  ↓
binary  1 = 0 = 0 = 1
```

that is, 1101 Gray = 1001 binary.

This conversion process could be programmed and included in the software. Alternatively a hardwired logic circuit with EOR gates could be interfaced between the encoder output and the computer input.

4.3 ASCII CODE

A binary code is also used to represent alpha-numeric characters in the interchange of information between the constituent parts of a microcomputer system. The most commonly used code for digital communication links is the American Standard Code for Information Interchange normally referred to as ASCII (Askey). This is a 7-bit code which accommodates 128 definable characters. The decimal equivalents of the ASCII codes for a sample of alphanumeric characters are as follows:

'A' = 65, 'B' = 66, 'C' = 67, . . . 'Z' = 90
'a' = 97, 'b' = 98, 'c' = 99, . . . 'z' = 122
'0' = 48, '1' = 49, '2' = 50, . . . '9' = 57

The ASCII code for any keyboard character can be obtained simply as indicated below:

> For a BBC microcomputer
>> Type: PRINT GET ⟨CR⟩

On pressing the 'A' key in the capital mode, a value of 65 is printed to the screen which is the ASCII code, in decimal, for 'A'.

Alternatively, the hex values can be obtained by responding to:

> PRINT ∼GET ⟨CR⟩

On pressing the 'A' key, the above returns 41 which is the ASCII code, in hex, for 'A'.

Similarly, in IBM's BASICA, the equivalent commands are:

> PRINT ASC("A")   which returns 65
> and   PRINT HEX$ (ASC("A")) which returns 41

4.4 ELEMENTS OF BASIC

Many textbooks have been written on the BASIC language and it is not the intention here to compete with these works. It has already been assumed that the reader will be studying the BASIC programming language in parallel and, as a consequence, much of the detail has been omitted. Fuller particulars on the BASIC commands may be obtained from the various references given at the end of this chapter.

The first true digitial computers, developed in the late 1940s, were based on valve technology and were programmed directly in binary codes. This task was greatly simplified with the introduction of high level languages. The most popular language is BASIC, which is an acronym for *B*eginners' *A*ll-purpose *S*ymbolic *I*nstruction *C*ode, and was devised in the mid 1960s. The first BASICs were relatively primitive but today a number of powerful dialects are available. These can more than adequately accommodate various types of input/output, numerical calculations, text manipulation, file handling and screen graphics. The majority of BASICs are 'interpreted', which means that the conversion from high level to machine code (binary) is done line by line at each execution of the code by means of an interpreter program. One such BASIC is that used on the BBC microcomputer, referred to as BBC BASIC. A number of textbooks, see references, have been devoted to this language.

BASIC on the IBM-PC is provided through BASICA and IBM compatibles might use PBASIC, GWBASIC or XBASIC. These latter versions of the language are all very similar and they are essentially compatible with IBM's BASICA in that programs written in these versions will all run on an IBM machine. Note however that the converse is not true and programs written in BASICA might not run on an IBM compatible machine.

When the BBC microcomputer is switched on it is ready to be programmed directly in BASIC but various operating system commands can be activated. The operating system used with the model B is the Disc Filing System (DFS) and that with the Master is the Advanced Disc Filing System (ADFS). The operating system allows the operator to SAVE, LOAD, CHAIN, DELETE and COPY files along with the ability to list a DIRectory of files and discs, or get INFOrmation about any specified file.

On IBM machines and compatibles, the operating system is PC-DOS, or the very similar MS-DOS. On start up, the machine will default to DOS and a BASIC program must first of all be created on a text editor before it can be saved as a BASIC source code. Fuller details of BASIC on 16-bit machines and MS-DOS are covered in chapter 8.

In the development of a program, various operational facilities are required of the language. The most important are summarised in BASIC syntax for input/ output, arithmetic and logic operations, conditional statements, looping, external functions and procedures. Input and output, disc file handling and graphics are covered in chapter 9.

(a) Arithmetic and Logic Operations

In addition to the usual arithmetic operators for addition, subtraction, multiplication, division and raising to a power, integer division (DIV) and integer remainder (MOD) operations can be performed. These are particularly useful for grouping decimal numbers into their equivalent high and low bytes in a 16-bit binary representation.

For example, using the example from section 4.1, to accommodate the number 50395 as two bytes, the low and high order byte values can be obtained as follows:

Low byte = 50395 MOD 256   High byte = 50395 DIV 256
           which returns 219   which returns 196

The logical AND and OR operators are useful in measurement and control operations for reading the state of a bit, setting a bit to '1' or clearing a bit to '0' in any particular byte.

To read the state of a bit:
The byte is say 110X 0111 and the state of X is required.
A mask of 0s is chosen except for the bit to be read.
That is, mask = 0001 0000, then perform a logical AND.

    byte is   110X 0110
    AND with mask   0001 0000
                    _____
    result   000? 0000

If the result is zero then X is '0', otherwise X is '1'.

To set a bit to '1', perform a logical OR with a mask of 0s, except for the bit to be set.

Say bit X is to be set to logic '1' in a byte at state 110X 0110

OR with mask	0001 0000
result	1101 0110

This sets bit 4 but retains the condition of all the other bits.

To clear a bit to '0', perform a logical AND with a mask of 1's, except for the bit to be cleared.

Say bit X is to be cleared in a byte at state 110X 0110

AND with mask	1110 1111
result	1100 0110

This clears bit 4, but retains the condition of all the other bits.

(b) Conditional Statements

The 'IF ⟨statement⟩ THEN ⟨statement⟩ ELSE ⟨statement⟩' instruction sequence in a program is used in conjunction with the relational or logical operators to give conditional branching. For example:

```
        100 IF temp = limit THEN GOSUB 800
or      200 IF(L*D)⟨MAX THEN X=Y ELSE X=Z
or      300 IF (I AND 16)⟨ ⟩0 GOTO 300
```

(c) Looping

Programs can be made to do repetitive operations and a number of methods are available in BASIC for carrying out this task. The most common method is to use a FOR/NEXT loop.

```
        FOR K=1 to N STEP 1
        ⟨statements in loop⟩
        NEXT K
```

The step increment need not be specified if it is unity.

FOR/NEXT loops can conveniently be used with arrays where a number of similar data types are to be independently stored and processed.

A typical example is to read and store a number of data elements:

```
        10 INPUT "Number of tests made = "; N
        20 DIM PRESS(N), TEMP(N)
        30 FOR K=1 TO N
        40 READ PRESS(K), TEMP(K)
        50 NEXT K
```

The pressure and temperature values would be held as array elements.

An alternative to the FOR/NEXT loop is to use a counter checking method which involves:

starting the count with   C=0
incrementing the count with C=C+1
terminating the count with   IF C<N GOTO a specified line number prior
to the increment instruction

A useful method of looping, peculiar to BBC BASIC, is the REPEAT/UNTIL facility. This causes the computer to repeat a set of instructions until some prescribed condition is met.

REPEAT
⟨statements in repeat loop⟩
UNTIL temperature ⟩= 100

The essentially equivalent structure in IBM BASICA is provided with the WHILE/WEND commands. For example:

WHILE (temperature ⟨=100)
⟨statements in repeat loop⟩
WEND

(d) External Functions and Procedures

An external function is a section of a complete program which culminates in the production of a specific value from a statement in the main program. An illustrative application of this is the correlation formula relating the pipe friction coefficient (f) to the Reynolds Number (Re) and pipe relative roughness (k/D) in hydraulics.

$$f = \left[\left(\frac{8}{Re}\right)^{12} + \left(\frac{1}{(A+B)^{3/2}}\right)\right]^{1/12} \times 0.3557$$

$$\text{where } A = \left[2.457 \ln \left\{ \frac{1}{\left[\left(\frac{7}{Re}\right)^{0.9} + 0.27\left(\frac{k}{D}\right)\right]} \right\} \right]^{16} \times 10^{-6}$$

$$\text{and } B = \left(\frac{37530}{Re}\right)^{16} \times 10^{-6}$$

This correlation simulates the well known pipe friction chart and enables f to be evaluated within a computer program from specified values of Re and k/D. The statement in the main program each time this is required is simply:

F = FNfunc(Re,KD)

The external function, 'func', is defined at the end of the main program as follows:

```
280 REM f=func(RE,k/D)
290 REM using CHURCHILLs CORRELATION
300 DEF FNfunc(RE,KD)
310 LOCAL C1,C2,C3,C4,C5,C6,C7,A,B
320 C1=(7/RE)^0.9
330 C2=C1+(0.27*KD)
340 C3=1/C2
350 C4=LN(C3)
360 A=((2.457*C4)^16)/1E6
370 B=((37530/RE)^16)/1E6
380 C5=(8/RE)^12
390 C6=((A+B)^(3/2))
400 C7=1/C6
410 FUN=((C5+C7)^(1/12))*0.3557
420 =FUN
430
440 REM*****************************
```

The above function is written in BBC BASIC. In Borland's TURBO-BASIC the statements would take exactly the same form as shown above except for the last line, 420, where the function would be terminated with END DEF. In IBM's BASICA however, the function definition, DEF FN, has to fit into a single line of coding. After the function has been defined it can then be used to perform calculations on its argument.

In contrast, a procedure or subroutine is a section linked into a complete program. A procedure may specify a way of performing tasks and it is called from within a main program. A simple illustrative example is a procedure to produce a specified time delay within a BBC BASIC program.

```
10 CLS
20 INPUT"Delay required in seconds= ";seconds
30 PROCdelay(seconds)
40 PRINT "Delay time complete!!!"
50 END
60 REM ****************************
70 DEFPROCdelay(S)
80 TIME=0:REPEAT:UNTIL TIME=S*100
90 ENDPROC
```

It should be noted that the input parameter, 'seconds', in the main program equates to the dummy parameter, 'S' in the defined procedure. It is usually unnecessary to specify parameters in this way and it is often preferable to use variable declarations which are common to both the main program and the procedure. These are referred to as 'global' parameters.

Procedures are not available in IBM's BASICA. They are available in various forms however, in some of the other languages similar to BASICA. The same essential features can be implemented nonetheless, with the GOSUB instruction.

To use the GOSUB facility, the command is simply followed by the line number in the program to which control is to be transferred.

The listing for the above procedure to produce a specified time delay in a BASICA program is as follows:

```
10 CLS
20 INPUT "Delay required in seconds= ";seconds
30 GOSUB 70
40 PRINT "Delay time complete!!!"
50 END
60 REM ****************************
70 REM Delay routine
80 T=TIMER:WHILE (TIMER-T)<seconds:WEND
90 RETURN
```

The command RETURN in the last line of the subroutine transfers control back to the line immediately following the GOSUB statement.

4.5 FLOW DIAGRAMS

A particular means of describing an algorithm is to use a flowchart which displays discrete events. This enables the logical processes within the program to be viewed in a much more coherent manner than a program listing. Standard flow charting symbols are used and these consist of three basic types, one for start/ stop, one for an operation or assignment and one for a decision. The instructions inserted into the boxes should be independent of the programming language used to develop the software.

Example

A computer-controlled central heating system has two temperature sensors, one for the room temperature and one for the water temperature in the boiler. Outputs from the computer control the on/off functions relating to the boiler and the water circulating pump. The control strategy to be adopted would be as follows:
(i) If the room air temperature is higher than the temperature setting, then both pump and boiler are to be switched off.
(ii) If the water temperature is higher than its temperature setting, then the boiler should be switched off but the pump should be left on.

The corresponding flow diagram, from which the computer program could easily be developed, is given in figure 4.1.

There are of course other methods of organising the input/output in this application, but the flow diagram is illustrative of its use in developing the computer program to perform a specific task.

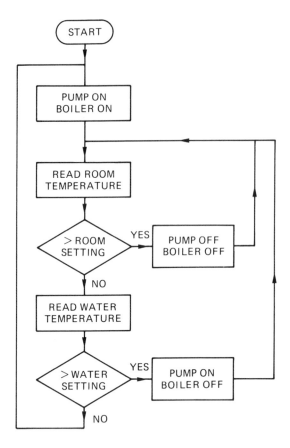

Figure 4.1   *Control strategy for a central heating system*

4.6   ELEMENTS OF ASSEMBLY LANGUAGE

BASIC is a high level language and the conversion to machine code through the interpreter program takes time. When operating speed is important, or when machine codes are required for insertion into an EPROM, see section 5.2, a low level language must be used.

Low level programming of a microprocessor, in machine code, involves the writing of a program by sequentially specifying the required instructional codes in binary or hexadecimal form appropriate to the processor used. These codes, inserted into the computer memory, relate to such operations as:

    Arithmetic   − add and subtract
    Logic   − AND and OR
    Data transfer − load and store

Shift program control - jump to new address to another area of memory

For example, with the 6502 microprocessor, the command 'jump to the hex address 24ED' would have the hex machine code instructions:

4C ED 24

The command is identified by the hex number 4C and is followed by the relevant 16-bit address with the lower byte given first.

Assembly language programming involves writing the program using mnemonic instructions which equate on a one-to-one basis to the machine code form. This makes the low level programming easier and an assembler program can then be used to translate the mnemonics into the appropriate code for the processor used. Ultimately the code is stored into memory in binary form. The binary code however is virtually unreadable since it simply consists of groups of 1s and 0s.

For instance, the above example would be written in assembly language as:

JMP   24ED

Compare this with the actual binary code which would be stored in memory:

0100 1100   1110 1101   0010 0100

All assemblers operate on substantially the same principle, which is to translate the assembly language instructions into machine code and place the code in a specified area of memory. The 8-bit 6502 microprocessor, with its proven performance and ease of coding, has been adopted by many manufacturers of computer-based data-acquisition and control systems. The 6502 is a 40 pin IC which provides the following facilities:

56 instructions
12 addressing modes
8-bit data bus and 16-bit address bus
1 or 2 MHz system clock frequency
Interrupt facilities and stack operations

The registers in the 6502 are as follows:

ACCUMULATOR	8	A
INDEX REG. X	8	X
INDEX REG. Y	8	Y
PROGRAM COUNTER	16	PC
STACK POINTER (1)	8 + 1	SP

The function of each is:

Accumulator   data transfer is carried out via this register and the results of arithmetic and logic operations are placed in it.

Program Counter holds the address of the next instruction to be operated on by the CPU during execution of a program.

During each execution the PC increments.

Index Counters   used in addressing modes involving indexed addressing.

They are specifically useful for accessing successive elements of tabulated data.

Stack Pointer   the 'stack' is an area of memory, confined to page 1 for the 6502, and used for the storage of data such as the address that the PC must return to after the execution of a subroutine. The SP holds the address of the current position of the stack.

Another important register is the processor status register. This contains FLAGS that are set (logic '1'), or cleared (logic '0'), depending upon the nature of the data currently stored in the accumulator, or index registers, as a result of any arithmetical or logic operations made by the processor. Other flags are altered as a result of interrupts. The flags in the processor status register are as follows:

7							0
sign N	overflow V	_	break B	decimal D	interrupt I	zero Z	carry C

N negative flag; set if the most recent operation performed in the arithmetic and logic unit gave a negative result.

V overflow flag; set if the result of a signed number arithmetic operation is too large. Occurs when there is a carry from bit 6 into bit 7 with no external carry (i.e. bit 7 into bit 8), or when there is an external carry but no carry from bit 6 into bit 7.

B break flag; set by the microprocessor when an interrupt is caused by a break command.

D decimal flag; when set, arithmetical operations are performed in binary coded decimal format. When cleared, the operation is in binary (hex). Hex operation is usual.

I interrupt disable flag; when set, the interrupt request, IRQ, input will not interrupt the microprocessor.

Z zero flag; set by the microprocessor if the operation in the arithmetic and logic unit gave a zero result.

C carry flag; set when there is an overflow from an 8-bit arithmetic operation, i.e. from bit 7 into invisible bit 8.

Logical decisions from within a program, in the context of a measurement and control application, are normally based on the Z, C, and N flags in order of importance.

Addressing Modes

Addressing generally refers to the specification, within an instruction, of the location of the operand on which the instruction will operate. A different code is used with each addressing mode for a specific instruction. The most common instruction used in 6502 assembly language programming is LDA, which transfers data from a specified source into the accumulator. This instruction can be used to illustrate the main addressing modes.

(i) Immediate

This is a two byte instruction with the data (preceded by a #) immediately following the opcode. For example:

LDA#&FF, which in machine code is A9 FF

(ii) Absolute

This is a simple reference to a 2-byte address. For example:

LDA &0200   (AD 00 02 in machine code)

Note: the low byte (00) is read first in the machine code version.

(iii) Zero Page

This is the simplest form of addressing and refers solely to zero page locations (0000–00FF, i.e. 256 bytes per page). For example:

LDA &20   (A5 20 in machine code)

This loads the accumulator with data held at 0020.

(iv) Indexed Addressing

This is a technique specifically useful for accessing the data elements of a block or a table. The contents of X and Y are added to the address to give the actual reference address. For example:

if X holds data of &50 then LDA &0200,X (equivalent to BD 00 20) loads the accumulator from location &0250

A zero page mode indexed by X or Y also exists.

(v) Implied

This is a single byte instruction which makes no reference to an address. Neither does it involve data. Examples are BRK, CLC, SED, INC, NOP, RTS, TAX, etc.

(iv) Relative

This is used with the branch instructions and displaces the program counter from its current position if the branch is taken. Note: the branch is actually made from two byte locations beyond the location of the branch instruction. This is because the program counter always holds the address of the next instruction to be executed and these are all stored as 16-bit data. For example:

```
AGAIN DEY
      BNE  AGAIN
```

If the zero flag Z is not set, then the program counter is returned to the line signified by the label instruction, i.e. effectively 3 lines back in the example given above.

Note: 3 = 0000 0011 and −3 = 1111 1101 = FD. The machine code for the above branch instruction would be 88 D0 FD.

The branch instructions are taken conditional of the state of the flags in the status register and are analogous to the IF/THEN statement in a high level language. Relative addressing is used and the offset immediately follows the branch mnemonic. This offset is limited to a maximum of 127 locations, 7F hex, forward or 128, 80 hex, backward.

If a movement of greater than +127 or −128 is required then a jump instruction, JMP, must be used in conjunction with the branch statement.

A summary of the 6502 instruction set and the corresponding hex codes for the most common addressing modes are given in appendix II.

Example on 8-bit Addition

Say that the two numbers to be added are stored at zero page addresses of 0070 and 0071 hex, and that the result is to be stored at address 3000 hex.

```
CLD        \ set hex mode
LDA &70    \ load data from 0070 into accumulator
CLC        \ clear carry flag, C = 0
ADC &71    \ add with carry, contents of 0071 to accumulator
STA &3000  \ store result at address 3000
```

The carry flag keeps a check on a carry-over from bit 7 into the invisible bit 8. It must obviously be cleared prior to 8-bit addition, since the result of an ADC operation is to add the carry bit to the contents of the accumulator as well as the memory data. The carry bit is extremely useful in multi-byte arithmetic.

Conditional statements in low level programming, equivalent to 'IF ⟨statement⟩ THEN ⟨decision⟩' in BASIC, are useful in measurement and control application software. These instructions translate a comparison between data values, CMP, and a branching decision dependent upon a change of flag state.

The 6502 compare instructions are as follows:

CMP (A-M) compare with accumulator
CPX (X-M) compare with X register
CPY (Y-M) compare with Y register

The corresponding Z, C and N flag changes, as a consequence of a CMP instruction are shown in table 4.2.

Table 4.2

Condition	Effect on flag		
	Z	C	N
A < M	0	0	1
A = M	1	1	0
A > M	0	1	0

It is noted that the C flag is set for both the > and = conditions and hence a test for a strictly 'greater than' condition must use two instructions involving the Z and N flags to test for $Z = 0$ and $N = 0$.

Example on producing a single delay loop by decrementing a counter:

Address field	Label field	Operator field	Operand field	Comment field
A00	DELAY	LDA	#&FF	; counter = 255
A02		STA	&70	; store at 0070
A04	LOOP	DEC	&70	; decrement counter
A06		BNE	LOOP	; decrement until Z = 1
A08		RTS		; return from subroutine

The above example is displayed as a subroutine which is called from within the main program by:

JSR DELAY

The above is also illustrative of the use of labels in assembly language programming. Labels are handled by the assembler to calculate the offset for the relative addressing branch instruction.

The use of the logical operators, AND and OR, in low level programming is identical to that in high level as illustrated in section 4.4.

For example, to wait in a program until bit 4 of the byte at address FE60 goes low:

```
TEST  LDA  &FE60    \ read byte to be tested
      AND  &10      \ AND with mask 0001 0000
      BNE  TEST     \ repeat until Z = 1
```

On satisfying the Z flag condition, the program would then continue. In many measurement and control systems it may be unnecessary to develop software in a low level language. Consideration should always be given to the specific details of the actual application before making the decision on the choice of language. A combination of both is often the best alternative. A high level language is often used, for instance, to input fixed parameters at the keyboard and to display the results. A machine code routine may subsequently be CALLed from the main program, as necessary, when speed is important. Often the choice, or availability, of the hardware decides whether the software must be developed in low or high level. Such is the case for single chip or single card computers with EPROM firmware.

4.7 PROGRAM DEVELOPMENT AND STRUCTURE

Software development using a personal computer, with a high level language, requires little or no additional hardware other than the usual keyboard, display monitor, disc drives and printer. Additional development aids such as editor, assembler, disassembler, linker and EPROM emulator are essential when dealing with single chip and single card computers. These however are beyond the intended scope of this text.

The operating system ROM in the BBC microcomputer contains a 6502 assembler which is unusual in that it is called from within a BASIC program. This makes it easy to jump back and forth from machine code routines to the BASIC language. The problem resolves to one of placing the machine code into a 'safe' area of RAM, such that it is not overwritten by the interpreted BASIC. Various methods can be used, but a convenient storage area for the code is to place it into the RS423 serial communication port buffer. This area contains 256 bytes and starts at hex address A00. 256 bytes are usually sufficient for most basic illustrative measurement and control applications, but if this is not the case then another area of memory must be chosen.

The start and end of the assembly language program is indicated by square brackets [and], and the last instruction must be RTS — 'return from subroutine', which returns control to BASIC. The instruction 'RUN' will assemble the program and execution is performed by 'CALL' followed by the starting address.

The assembler is a 'one pass' type meaning that the assembly of the program is only done once and any branching or jumping forward to a specified label cannot be interpreted and an error message is displayed. This is easily overcome however, by first suppressing errors, and then reporting errors in a two pass loop.

As an illustrative example consider the exercise in section 4.6 on 8-bit addition.

```
 10 FOR pass -1 TO 3 STEP 2
 20 P%=&A00
 30 [
 40 OPT pass
 50 .start
 60 CLD          \set hex mode
 70 LDA &70      \load data from 0070 into accumulator
 80 CLC          \flag C-0
 90 ADC &71      \add contents of 0071 to accumulator data
100 STA &3000    \store answer at address 3000
110 RTS
120 ]
130 NEXT pass
140
150 INPUT "first number- ";FIRST
160 INPUT "second number- ";SECOND
170 ?&70=FIRST:?&71=SECOND
180 CALL start
190 PRINT "Answer- ";?&3000
200 END
```

P% is the assembler directive and in this case will start placing the assembled code from address A00. The CALL instruction then executes the machine code component of the program. However, once the machine code routine has been debugged, the code can be saved to a disc file with:

*SAVE CODE A00+FF ⟨CR⟩

This will save the whole page of memory, whether required or not, as a disc file named 'CODE'. The stored data codes can be displayed with:

*DUMP CODE ⟨CR⟩

For the example given, this will be:

D8 A5 70 18 65 71 8D 00 30 60

All that is now required is to ensure that the machine code program is loaded, then to CALL it up appropriately from within the BASIC program. It can be automatically loaded at the first line in the program with:

10 IF !&A00 ⟨ ⟩ &1870A5D8 THEN *LOAD CODE

Note that the number which &A00 is compared against must be the first four bytes of the machine code program, written in reverse order.

CALL &A00 will execute the program.

The above procedure for incorporating low level routines into a high level program, although specifically outlined for a BBC microcomputer, is a general presentation of the concepts associated with any high level computer system. See chapter 8 on how this is done on an IBM-PC.

Structure

Prior to developing the program for any data-acquisition or control exercise, it is essential to have a complete understanding and identification of the problem such that the software can be constructed according to a defined systematic framework. This involves a formal approach to the structuring of the program and there are two basic techniques which can be applied.

(1)   The 'top-down' approach

This method starts with the problem definition and the program statements follow the order of operations to be executed giving a logical top-down sequence of instructions. It is relatively quick to generate software but the technique has the disadvantage of making lengthy programs difficult to understand. Software generation, in addition, tends to be an individual effort since the program is one large entity.

(2)   The 'modular' approach

The problem is first divided into sub-tasks that represent known solutions which can be independently translated into program code. These are then combined into larger pieces until a complete solution is achieved through a series of sub-routines or procedures. This incorporates the benefits associated with the collection of standard modules, for use in different applications, and also makes any debugging or alterations easier to perform. This technique nonetheless, makes software development more expensive and requires good documentation.

In general, the structuring of a program to fulfill a specified task is basically an iterative procedure and a combination of both approaches is normally adopted.

EXERCISES

1. Calculate the positive and negative range of integer numbers which can be represented on three bytes of data using a two's complement notation.
   [−8 388 608 to +8 388 607]
2. Convert the Gray Code value of 0011 0001 into a hexadecimal number.
   [21]
3. Outline the relative merits of high and low language programming for a data-acquisition and control application.
4. Determine the effect of the Z, C and N flags when the following 8-bit operations are executed in a 6502 based assembly language program:

   (i)   120-25 (Z = 0, C = 1, N = 0)
   (ii)   52-52 (Z = 1, C = 1, N = 0)
   (iii)   8-23 (Z = 0, C = 0, N = 1).

5. A conversion from an analogue voltage level to a corresponding numerical value is initiated on an analogue-to-digital converter chip by pulsing (logic '0' to logic '1') the 'start conversion' pin on the IC. When the conversion is complete, the 'end of conversion' pin on the IC indicates a logic change from '1' to '0'. At this stage the converted value can be read from a specified input port. Draw the flow diagram for this process which could be implemented as a subroutine.

6. Convert the flow diagram, given in the example illustrated in section 4.5, into:
   (i) a BASIC program
   (ii) an assembly language program.

7. A microswitch is connected to bit 2 of an input port, addressed at FE60, in such a manner that each time it is activated the bit goes low. Write a program in BASIC and also one in assembly language which would set bit 6 of an output port, addressed at FE61, permanently to a logic '0' as soon as the switch is depressed.
Repeat for the case which would set output bit 6 low when the switch is on, but reverts to a high state as the switch is put off.

8. The figure shows the square wave pulse trains, A and B, from an incremental encoder which are fed into bits 0 and 1 of an 8-bit microprocessor port. These indicate that for clockwise rotation signal A leads signal B with the opposite occurring for the anticlockwise case.
Draw up a flow diagram and write a 6502 assembly language program which would distinguish between clockwise and anti-clockwise rotation using an available subroutine at F000 to display 'clockwise' and one at F200 to display 'anti-clockwise' on a screen monitor.

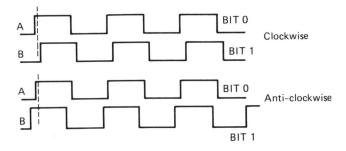

REFERENCES

Borland International Ltd, *TurboBasic, IBM version*, Borland International Ltd, 1987

Bright, B., *Pocket Guide – Assembly Language for the 6502*, Pitman, 1983

Meta(i), *Development System Based on BBC and IBM Microcomputers*, Crash Barrier, 23 Senwick Drive, Wellingborough, Northants, NN8 1RY

Ross, C. T. F., *BASIC Programming on the AMSTRAD 1512 and 1640, and IBM Compatibles*, Jonathan Ross, 1987

White, M. A., *Good BASIC programming on the BBC Microcomputer,* Macmillan. 1984

Zaks, R., *Programming the 6502*, Sybex, 1980

Chapter 5
Microprocessor Technology

5.1 SYSTEM ARCHITECTURE

Since its inception in the mid-1970s there are now currently a number of different microprocessor designs available in several versions. Nearly all the 'popular' designs are produced by a few manufacturers and these include such companies as Intel, Motorola, Rockwell, Texas Instruments and Zilog.

Microprocessor-based systems require additional family support chips and, in true digital form, all microelectronic components which constitute a microcomputer are designated numerically rather than by name. Some of the more popular microprocessors available are as follows:

(a) *8-bit* Binary data handled in a 'word', 8-bits wide, defining the accuracy of the number handling representation. For example:

> Intel 8080 and 8085
> Motorola 6800 series
> Rockwell 6502 series
> Zilog Z80

(b) *16-bit* Binary data handled in a 'word' of 16-bits width. For example:

> Intel 8086
> Motorola 68000
> Zilog Z8000

The 8086 is one of the most powerful and versatile 16-bit microprocessors available and it has been widely adopted by industry. Further enhancements include the 32-bit versions, 80286 and 80386, which are used in the IBM PS2 microcomputers to provide increased processing power.

With the development of the microprocessor, there has been an increasing integration of microelectronics within a wide range of industrial and scientific environments. The most significant developments are associated within the fields of instrumentation and measurement. Although it is unnecessary, for the user of the technology, to understand in detail how each individual chip actually functions, it becomes essential to have at least a working knowledge of the logical organisation of the system hardware and how each component relates to each other. The composition of this hardware structure is known as the system architecture.

A digital computer system comprises three main constituent parts: these being the microprocessor, the memory and the input/output. Digital signals which have a common function are transmitted between the main components by a group of wires, or conduction tracks, termed a bus. In a microcomputer, there are three buses, i.e. the data bus, the address bus and the control bus.

The interconnection between the basic hardware components in a microcomputer is illustrated in figure 5.1.

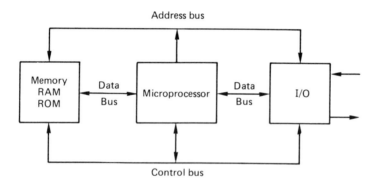

Figure 5.1 Basic components in a microcomputer

The microprocessor is a very-large-scale integrated circuit (VLSI), which is the brain of the microcomputer system and acts as the central processing unit (CPU). ICs are generally classified according to the number of components on the silicon chip and VLSI has tens of thousands. The Intel 8086 microprocessor, introduced in 1979, has 29,000 transistors packed on to the 225 mm^2 chip.

The main feature of the microprocessor is the arithmetic and logic unit (ALU). The ALU allows the arithmetical manipulation of data, addition and subtraction with 8-bit systems and multiplication with 16-bit systems. Logical operations (AND, OR, etc.) can also be performed. In addition to the ALU, the CPU contains a number of temporary data storage registers to fetch, decode and execute instructions, and to time and control circuits to manage the internal and external operation of the complete microcomputer system.

The processing power of the CPU is influenced by such factors as word length, instruction set, addressing modes, number of available registers and information transfer rates. For word processors, or the manipulation of large quantities of data as in CAD packages, 16- or 32-bit microprocessors are essential. In the field of measurement and control, 8-bit systems are usually adequate. It should be noted that the instruction sets used for preparing the application program, in either assembly language or machine code, are different for each type of microprocessor. Compatibility does exist to some extent though, between the Intel devices and that of Zilog.

The system clock, accurately controlled by a quartz crystal maintaining a constant frequency, acts as the heart beat for the system and provides all of the timing reference points.

All the basic components, CPU, memory and I/O, and their interconnections, may reside in a complete microcomputer system encompassing keyboard, monitor, etc. Alternatively, they may reside on a single card, or even on a single chip to give a single chip microcomputer.

5.2 MEMORY DEVICES

Memory units consist of those used to store binary data, which represents the user program instructions, and those which are necessary for the user to operate the system.

Memory takes the form of one or more integrated circuits. These basically hold locations capable of storing a binary word. Each location is assigned a unique address within the system, and data can be selected through the address bus. As a binary code is deposited by the CPU on the address bus, defining a specific location in memory, the contents of that location are selected and placed on the data bus. The appropriate piece of memory hardware and specific location is selected by means of an address decoding circuit built up from logic gates within the microcomputer system. The end result is a highly flexible data manipulation arrangement.

In an 8-bit microcomputer, i.e. 8-bit data bus, the address bus is 16 bits wide. This enables 2^{16} = 65536 locations to be addressed, and specifies the total memory capacity of the machine as 64K (i.e. $2^6 \times 2^{10}$ = 64×2^{10}, and 2^{10} is 1K byte).

Thus, the memory locations available span the addresses:

0000 0000 0000 0000 − 1111 1111 1111 1111 in binary
   0   0   0   0 − F   F   F   F   in hex

The memory is further subdivided into pages with the high order byte of the address denoting the page number and the low order byte indicating one of the 256 locations available on each page. For example:

OE FF denotes the last byte on page E

In PCs with 8086/8088 microprocessors, an additional 4 bits are effectively made available on the address bus. This theoretically constitutes one megabyte of addressable memory. With the IBM PS2 microcomputers, employing the 80286 and 80386 microprocessors, the address bus is 24-bits wide and can address up to 16M bytes of physical memory. This releases new levels of processing power to accelerate the processing speed in measurement and control applications. In any event, data storage within the system is effectively the sequential holding of binary words at uniquely specified addresses.

The types of memory chips built into the system basically divide into two categories:

(i) *Random Access Memory (RAM)* — where data can be read from or written to any specified location.

RAM is more correctly defined as read/write memory, and data retention is dependent upon power being applied to the device. This type of memory is normally employed for the temporary storage of the computer programs, at the editing or execution stage, or the storage of data from measuring transducers prior to permanent storage as a disc file. In a number of measurement systems available, the RAM is made non-volatile by providing battery back-up which is an extremely useful facility in data logging applications. Once the data has been acquired in the field and temporarily stored in battery backed-up RAM, it can then be transmitted from the microprocessor-based data logger to a microcomputer for display, processing or a more permanent form of storage.

(ii) *Read Only Memory* (ROM) — where data is held in a secure manner and can be read in any specified sequence.

Once the chip is configured it cannot be overwritten and the programs which specify the system operation, termed the monitor program, are 'burnt' into ROM when they are known to operate in a satisfactory manner. Basic ROM is inflexible since the contained software is developed by the system manufacturer. It is often useful however, to have all programs which are to be permanently stored in the microcomputer in a non-volatile form, held in an Erasable and Programmable Read Only Memory (EPROM). This is undoubtedly the most popular type of ROM used because the write process is reversible. These chips come in popular memory capacities of 2K, 4K, 8K, 16K and 32K and they are respectively designated by name as 2716, 2732, 2764, 27128 and 27256. The numbers following the '27' indicate the number of kilo-bits of memory available within the device.

EPROMs are supplied in an uncommitted form with each location holding FF hex. They are configured using an EPROM programmer which 'burns' or 'blows' the required data, in machine code form, onto the chip. If an error in the data exists, or an alteration is to be made, then the complete EPROM can be returned to its uncommitted state by exposing the small 'window' in the device to intense ultra-violet light for about 20-30 minutes. EPROM erasers are available for this purpose. Once programmed as required, it is usual to cover the window with opaque material. If uncovered, it would normally take some months before program corruption was experienced through the effects of natural sunlight.

A similar type of memory device is an Electrically Erasable Read Only Memory (EEPROM or E^2PROM). This is essentially similar to the EPROM but enables the user to alter any particular byte of data rather than wiping the entire chip. E^2PROM is not so popular as the EPROM, probably because of economic reasons.

5.3 I/O STRUCTURE

With the microprocessor acting as the brain of the microcomputer system and the memory chips storing the system operating software and application pro- grams, the other essential hardware required is that associated with the input and output of data in essentially binary form. Interface support chips associated with the various microprocessor families are available to enable communication with such hardware essentials as keyboards, display monitors, disc drives and printers.

The same I/O interface circuits are used in measurement and control applica- tions and the main functions required of the devices are:
1. Digital I/O logic lines which can be read or set by the microprocessor.
2. Data direction register to configure lines as either input or output.
3. Handshake lines to supervise data transfer via the I/O lines.
4. Timing and counting facilities.

The software used for controlling the communication between the micro- computer and other external devices is dependent upon the I/O interfacing technique employed. The two most common methods are either 'memory mapped', or 'dedicated port addressed'.

Memory Mapped I/O

In this method the I/O chip is connected into the system in the same way as the memory illustrated in figure 5.1. The I/O lines are contained as groups of 8-bits termed a 'port' and this byte is addressed in the same manner as any other location in memory. The port is accessed using memory transfer instructions like PEEK and POKE in high level BASIC, or LDA and STA in low level 6502 assembly language (see section 6.1).

Since the interface is connected into the bus structure, in exactly the same way as the RAM and ROM, no additional decoding hardware is required. Memory addresses are however, used up for I/O and as a result, communication is slower than the port addressed alternative.

Dedicated Port Addressed I/O

This method involves a second dedicated I/O data bus as shown in figure 5.2.

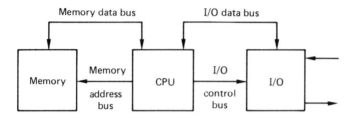

Figure 5.2   Port addressed I/O

When data is to be input or output, the necessary control signals are sent from the CPU to the I/O interface chip and the port data is transmitted via the dedicated I/O data bus. This does not effect the addressing of memory within the system and results in faster data transfer than with the memory mapped technique. The ports, or channels, are assigned unique addresses (numbers) on the dedicated bus and are accessed using the additional software instructions of IN (or INP) and OUT in both low and high level programming languages (see sections 8.4 and 8.5).

Although a number of I/O support chips are available, there are essentially two which figure prominently with the memory mapped and port addressed techniques. These are the 6522 versatile interface adapter (VIA), usually associated with the memory mapped 6502 microprocessor systems, and the 8255 programmable peripheral interface (PPI), associated with such processors as the 8080, Z80 and 8086 in port addressed systems.

The 6522 VIA

This is a general interface chip which provides such interface functions as two 8-bit parallel bi-directional ports, each with a pair of handshake lines, and two 16-bit counter timers.

The ports, often designated as data registers A and B ('DRA' and 'DRB'), each have an associated data direction register ('DDRA' and "DDRB'), which is used for setting a bit on a port as either an input or an output. A logic '1' placed in any bit of the DDR configures the corresponding bit of the port as an output. A logic '0' will configure it as an input.

The addresses follow the sequence PORTB, PORTA, DDRB and DDRA. For example, the program instructions to set the 4 MSBs of PORTB addressed at 0900 hex as output and the 4 LSBs as input are:

In BASIC:

POKE &0902,240

In 6502 assembly language:

LDA #&F0

STA &0902

If a.c. or d.c. loads such as solenoids, motors or lamps are to be driven from the port logic signal levels, then a power scaling interface, such as a Darlington driver, compatible with the microprocessor VIA must be used (see section 6.2).

The VIA control lines CA1, CA2, CB1, and CB2 can be set to operate in various read/write modes. This is achieved through the peripheral control register (PCR) in the VIA. CA1 has no output capability, but CA2, CB1 and CB2 can all be used as either input or output. These lines are incapable, however, of switching a power scaling device.

Active transition of the logic level on any of the control lines is 'flagged' in the interrupt flag register (IFR) and passed onto the microprocessor interrupt request (IRQ). This temporarily stops the processing of the current task until the interrupt has been serviced. The IRQ pin is level sensitive and when activated by a logic '0', the last address executed is stored on the stack to facilitate the return to the current program. The IRQ can be disabled if the interrupt disable flag in the processor status register is set. The flag is cleared, or set, by the assembly instructions CLI or SEI. An application using the VIA control lines to operate and monitor an A/D converter is given in section 7.2.

Another register within the VIA, associated with the enabling or disenabling of control line interrupts, is the interrupt enable register (IER), which defaults to all interrupts enabled.

The two programmable timers within the 6522 are generally referred to as T1 and T2. These are 16-bits wide and implemented as two 8-bit registers with a low byte/high byte arrangement. The modes of operation, selected by writing the appropriate code to the auxiliary control register (ACR) are:

1. Generate a single time interval.
2. Generate continuous time intervals (T1 only).
3. Produce a single or continuous pulses on bit 7 of DRB (T1 only).
4. Count high to low transitions on bit 6 of DRB (T2 only).

Bits 5, 6 and 7 of the ACR control the operation of the timers. A hardware generated delay using the T2 timer, in a single shot mode, can be programmed by making bit 5 of the ACR equal to 0, i.e. ACR5 = 0. The low and high order counter addresses, denoted T2C-L and T2C-H, are loaded in sequence and upon writing to T2C-H the 16-bit counter value decrements at the system clock rate. On reaching zero, bit 5 in the IFR, i.e. IFR5, is set and hence the end of the delay period can be detected. The T2 related interrupt flag, IFR5, can then be cleared by either reading T2C-L, or writing to T2C-H. This is best illustrated by means of an example on the controlling of the speed of a stepper motor.

A stepper motor is a power device which converts a d.c. voltage pulse train into a proportional mechanical movement. This rotates an output shaft a finite amount for each pulse received. A typical step angle is 7.5 degrees with the motor requiring 48 pulses/rev. The pulses are produced from a computer output port through a high-speed switching solid-state interface such as a Darlington driver. A four-phase stepper motor requires each phase to be pulsed in sequence, with each of the four steps moving the motor through the specified angle, e.g. 7.5 degrees. A typical sequence requiring four control bits is:

step	bit 1	bit 2	bit 3	bit 4
1	1	1	0	0
2	0	1	1	0
3	0	0	1	1
4	1	0	0	1

Reversing the sequence from step 4 to step 1 will reverse the direction of rotation of the motor. A solid-state integrated circuit, SAA1027, such as RS 300-237, is available which eliminates the need to pulse each phase. A single pulse (P) moves the motor one step, and a direction bit (D) controls the direction of rotation with a logic '0' denoting clockwise rotation and a logic '1' anti-clockwise. The pulse train need not be symmetrical, since it is the period which determines the rotational speed.

Consider the continuous driving of a 48 pulses/rev stepper motor with the SAA1027 interface, at a specified speed in rev/min and a specified direction, from a port on a microcomputer. The T2 hardware timer in the 6522 VIA is used to generate the necessary pulse lengths. The drive pulse (P) may be taken from the LSB of the port and the direction pulse (D) from bit 1.

If the pulses are sent as a symmetrical train, then the delay will be one half of the period which is related to the motor speed and number of pulses/rev. For example:

For a motor speed of N rev/min the period is
$$T = (60/48N) \times 1000 \text{ ms/pulse}$$
$$= 1250/N$$
Hence the required delay $= 0.5 \times T$
$$= (625/N) \text{ ms}$$

This delay is to be produced by the T2 hardware timer in the 6522VIA. The delay required in μs is $(625000/N)$.

The corresponding numerical value to give a speed ranging between $10 < N$ rev/ min < 400 must be stored as two bytes, say at zero page addresses &70 for the low byte and &71 for the high byte. Address &72 can be used to hold the condition for clockwise or anti-clockwise rotation.

An assembly language program to pulse the motor at the correct frequency, for a specified speed with declared register addresses is:

```
 10 DDR=&FE62:PORT=&FE60
 20 ACR=&FE6B:ifr=&FE6D
 30 T2CL=&FE68:T2CH=&FE69
 40 FOR PASS=1 TO 3 STEP 2
 50 P%=&A00
 60 [OPT PASS
 70 CLD
 80 LDA#03
 90 STA DDR        \ set port bits 0 and 1 as output
100 .repeat
110 LDA#01          \ send high part of the pulse
120 ADC &72         \ for specified direction
130 STA PORT
140 JSR delay
150 LDA#0           \ send low part of the pulse
160 ADC &72         \ for specified direction
170 STA PORT
180 JSR delay
190 JMP repeat      \ repeat continuously
200
210 .delay
220 LDA ACR         \ delay subroutine
230 AND#&DF         \ clear ACR5
240 STA ACR
250 LDA &70         \ load low byte
260 STA T2CL
270 LDA &71         \ load high byte
280 STA T2CH
290 .check          \ check for end of count
300 LDA ifr
310 AND#&20         \ ie IFR5=1
320 BEQ check
330 LDA T2CL        \ clear interrupt flag
340 RTS
350 ]
360 NEXT PASS
370 END
```

The delay counters to be stored at &70 and &71 can be evaluated from the specified speed in rev/min. The assembled machine code routine could then be called from within a BASIC program.

For example, say the code is stored in the RS423 buffer of a BBC microcomputer as illustrated in section 4.7. The corresponding program in BBC BASIC with the above machine code insert called 'DRIVE', is:

```
 10 IF !&A00<>&8D03A9D8 THEN *LOAD DRIVE
 20 INPUT"1.Motor Speed (10-400rev/min) ";N
 30 IF N>400 OR N<10 THEN 20
 40 INPUT"2.Required motor direction(CW/ACW)";D$
 50 Delay%=625000/N
 60 ?&71=Delay%DIV256:REM**High Byte**
 70 ?&70=Delay%MOD256:REM**Low Byte**
 80 IF D$="ACW" THEN ?&72=2 ELSE ?&72=0
 90 CALL &A00
100 END
```

For examples of other modes of operation of the timers, see the references at the end of this chapter.

The 8255PPI

All microprocessor families have parallel I/O interfaces designed for use with the particular type of CPU. The 8255PPI is used basically with Intel 8 and 16-bit devices such as the 8080 and 8086/8088.

The 8255PPI provides three 8-bit bi-directional ports which may be operated in three modes. No other functions such as timing or additional handshaking are available. The ports are designated as A, B and C and data direction is specified by writing to a write only Control Register.

The ports are addressed on the dedicated bus relative to a set base address within the system as follows:

	Base address
Port A	+0
Port B	+1
Port C	+2
Control Register	+3

The control word, to be written to the control register to set up the ports, is formulated as follows:

bit 7	mode set flag — active high	
bits 5 and 6	mode select (Port C upper nibble, Port A)	

bit	5	6
mode 0	0	0
mode 1	0	1
mode 2	1	X

bit 4	port A direction	0 = output, 1 = input
bit 3	port C upper nibble direction	0 = output, 1 = input
bit 2	mode select (Port C lower nibble, Port B)	0 = mode 0, 1 = mode 1
bit 1	port B direction	0 = output, 1 = input
bit 0	port C lower nibble direction	0 = output, 1 = input

The most common mode of operation is mode 0, which provides basic input and output operations for each of the three ports.

For example, the control word to make port A input and ports B and C output would be:

bit	7	6	5	4	3	2	1	0
	128	64	32	16	8	4	2	1
control word =	1	0	0	1	0	0	0	0

i.e.   144   or   90 hex

The corresponding instruction in BASIC, used with an IBM-PC type of microcomputer fitted with a digital I/O card having a base address set as 1B0 hex, is:

OUT &H1B3,&H90

Data can be output or input with:

OUT port address, value
I = INP (port address)

In 8086 assembly language programming on a PC, all I/O data transfer with addresses greater than 8 bits is performed using the DX register (see section 8.5), to store the I/O addresses. Output and input of port data is then made via the accumulator register, AX.

The appropriate mnemonic code instructions for setting up ports as previous is:

MOV DX,1B3H
MOV AL,90H
OUT DX,AL

Data is then output from the Port B with:

MOV DX,1B1H
MOV AL,value ; 'value' is the 8-bit value to output
OUT DX,AL

Data is read from the port A and transferred to the accumulator with:

MOV DX,1B0H
IN AL,DX

Further details on the above 8086/8088 assembly language instructions are given in section 8.5.

If hardware timing is required then a separate programmable counter/timer device must be used. One commonly adopted with an 8255PPI is the Intel 8253 chip which provides three independent 16-bit counters, each with a count rate of up to 2.6 MHz. The 8253 has various modes of operation and works basically on the same principle as the timers in the 6522VIA.

Although the majority of add-on cards for IBM-PC type machines are 8255 based, because of the popularity and versatility of the 6522VIA, cards with the 6522 are now generally available for use with PCs.

Direct Memory Access (DMA)

In data-acquisition systems involving analogue and digital signals suitably conditioned for inputting to a microcomputer, there is a limitation of about 100 kHz on the sampling rate when using direct program control to transfer data to memory.

If it was necessary to acquire the maximum amount of data at the highest speed, using the maximum amount of computer's resources, then the DMA technique might be employed.

This is a hardware technique which causes the microprocessor momentarily to abandon control of the system buses so that the DMA device can directly access the memory. The DMA controller, connected to the I/O interface, needs to know how many bytes are to be transferred, and where in memory the input data is to be stored. The data transfer rate is much faster than in an interrupt servicing method and data sampling rates of the order of 1 MHz are possible for most microcomputers.

5.4 BUS STRUCTURE

As outlined in section 5.1, the connection between the system components is made by an arrangement of three buses:

1. The *data bus* transmits the words in binary form representing the program instructions for the CPU to execute. It can also carry the information transmitted between the CPU and memory or I/O devices. Although the popular PCs use a 16-bit data bus, 8-bit data transfer operations remain the norm in data acquisition and control applications.

2. The *address bus* transmits the memory address related to the data bus information. In 8-bit systems this bus commonly has 16 lines to give 64K of addresses. PCs usually have an effective 20 lines to give 1M byte of available addresses, although software limitations often restrict this to 640K.

3. The *control bus* transmits the collection of timing and control signals which supervise the overall control of the system operation.

The physical format of a busing system is basically a circuit board with a number of connectors. Different types of microprocessors require different hardware interfaces and to alleviate the problems, standard bus structures have been developed in order to facilitate the connection of hardware components. In industrial type systems, cards for various microcomputer functions such as processor, memory, digital and analogue I/O, power switching and so on slot into a standard backplane or motherboard rack. This offers the advantage of being able to plug any specific card, designed to the bus standard, into a free slot in the rack to build up the system as required.

The physical form of the bus is represented by its mechanical and electrical characteristics. Such information as card dimensions, input and output pin-out connections, signal levels, loading capability and type of output gates must be known.

Standard buses are compatible with cards from different manufacturers and the most popular bus structures include Multibus, S-100 Bus, STD Bus, and the STE Bus.

Multibus

This is a bus structure developed for microcomputer systems manufactured by the Intel Corporation. It involves a relatively large printed circuit board which is 305 mm wide by 170 mm high. One of the two edge connectors on the card contains the general address, data, control and power lines. The bus, which has the ability to support more than one processor, provides for both 8-bit and 16-bit data transfers and can handle up to 24 address lines.

S-100 Bus

This is a popular standard bus structure, often associated with the hobbyist, and is so called from the number of pins used to implement the bus. Not all of the pins are used in most systems. The cards have dimensions of 254 mm wide by 130 mm high and the bus is normally associated with the 8-bit 8080 and Z80 microprocessors. The original design has been upgraded to provide for 16-bit data transfer and 24-bit addressing. It should continue to be an important standard for the future.

STD Bus

The modularity and simplicity offered by this well defined standard bus has led to the development of a large variety of board level products. These help provide innovative solutions to scientific and industrial applications and enable the simple interconnection of relatively small 8-bit data bus and 16-bit address bus systems. The STD bus also includes control lines for DMA operations. There are literally hundreds of I/O cards, dimensioned at 114 mm wide by 165 mm high, available to meet almost any interfacing requirement and thus providing a cost effective and flexible vehicle for a wide range of monitoring and control applications.

STE Bus

This is relatively new standard high performance bus structure. It has been designed to exploit the advantages of the new generation of microprocessors such as the developments based on the 8086. The board dimensions are relatively small at 100 mm wide by 160 mm high and 1M byte of available address space can support large high-speed data storage and manipulation tasks. Multiple processors can be supported on the same bus, with backplane frequencies up to 16 MHz. There are now an increasing number of manufacturers who produce cards to this particular bus structure.

PC Bus

Because of their excellent price/performance ratio and short user learning curves, personal computers are increasingly moving into industrial measurement and control applications. PC-based data-acquisition systems are now commercially competitive with the traditional larger computer systems.

There are basically two ways for the associated hardware to interface with the PC, i.e. via a standard serial or parallel communication-channel (see section 5.6) or a direct connection to the PC bus. The small size board level internal bus products are at the low end of the price range and typically occupy a single slot in the host computer. Cost is reduced because cards do not require their own separate enclosure or power supply, and high data transfer rates are achieved by eliminating the relatively slow external communications-channel protocol.

5.5 MEMORY MAP

The memory locations in RAM and ROM, which the processor can address, must accommodate space for such requirements as system monitor and utilities, user software and input/output. The manufacturer of the microcomputer assigns an area of memory for each functional requirement and provides the necessary information in a system memory map.

In 8-bit systems, with 64K of addressable memory, the memory map is usually composed of 32K of RAM and 32K of ROM or EPROM. The ROM holds the operating system software and normally some space is available in EPROM form for user firmware. In addition to providing space for user programs, the RAM area contains the system stack and the visual monitor data storage. The I/O facilities are also assigned an area of memory in a memory mapped system.

In a 6502- or 6800-based system, the RAM is usually low down in memory and the ROM is high up. A typical memory map is shown in figure 5.3.

The I/O is accommodated anywhere within the above structure and varies from one manufacturer to another. For example, the memory mapped input/output in the BBC microcomputer occupies three pages of memory FC00–FCFF, FD00–FDFF and FE00–FEFF, generally referred to as Fred, Jim and Sheila.

The display screen RAM is often fitted with a movable boundary in order to set variable resolution modes for graphics. Typically the screen memory need only be 1K for a text mode to give a 40×25 character display. This would be insufficient for computer graphics and a high resolution would require 20K of memory to MAP the screen. It should be noted that this greatly reduces the amount of RAM available to the user.

8080- and Z80-based systems have a distribution of memory similar to that as shown in figure 5.3. The ROM, however, is usually low down and the RAM high up in the memory map.

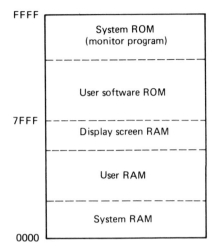

Figure 5.3 *System memory map*

The 16-bit PCs, with a 20-bit address bus, have 1M byte of addressable memory with the RAM comprising the first three quarters and ROM occupying the last quarter.

A general memory map, showing the distribution of RAM/ROM, for a PC is given in figure 5.4.

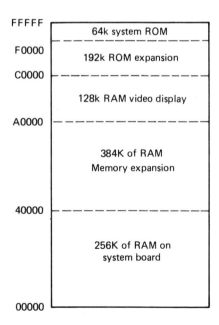

Figure 5.4 *General memory map for a PC*

Familiarity with the memory map of the system, to be used in any data-acquisition application, is essential since it indicates the areas reserved for the operating system. The programmer can then knowledgeably determine the locations available for data storage and machine code programs.

5.6 COMMUNICATION STANDARDS

Various standards have been drawn up to define the protocol for the transmission of binary data from within the microcomputer bus structure to external devices such as display monitors, printers and other peripheral equipment. Most microcomputers are equipped with this facility and manufacturers of data measurement and control instrumentation usually offer an external communication port as an extra.

The most commonly accepted standards are those defined by the American Electronic Industries Association (EIA) and the Institute of Electrical and Electronics Engineers (IEEE). The standards fall into the two categories of serial and parallel data communication. The difference between the two relates to the number of bits of information transmitted simultaneously between the devices. The serial method is the slower of the two, with the bits denoting the characters of information travelling sequentially along a single path. In the parallel method, the data word is sent as a parallel code, invariably 8-bits wide, resulting in a 'bit parallel, byte serial' transmission of information.

Serial Communication

Serial communication is the most common method used for the interconnection of a microcomputer to the relatively slow peripheral hardware, or between two computers, when transferring a low volume of information.

When communication takes place in a serial fashion, the 7-bit ASCII code is often extended to 8 bits. This is usually done by inserting a zero in the most significant position. In addition, a 'start bit', a 'parity bit' and one or two 'stop bits' are added. A start bit constitutes a high to low transition and informs the receiving device that a character code follows. The parity bit provides a check that no bits have been corrupted during transmission by ensuring that the sum of all the '1's in the ASCII group give either an even number for 'even parity', or an odd number for 'odd parity' setting. The stop bit, or bits, set at logic '1', terminate the transmission of a character and this is illustrated in figure 5.5 for the transmission of the letter 'A' with odd parity.

The transmission rate in bits/second is termed the 'baud' rate. A speed of 2400 baud corresponds to $2400/11 = 218$ characters/second since there are 11 bits required to transmit one character.

The (EIA) RS232C, or its successors the RS422 and RS423, is the most widely adopted standard employed and connection between devices is made via a standard 25-pin connector. This allows communication with one peripheral

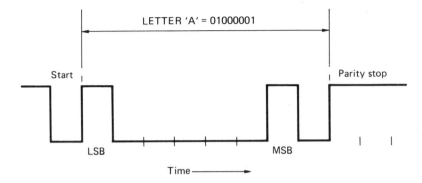

Figure 5.5 Serial transmission of the letter 'A'

device only. Twenty-one of the signal lines are defined in the standard although only five, or even three, are all that are usually required.

The three main connections are 'transmitted data' (pin 2), 'received data' (pin 3) and 'signal ground or common return' (pin 7). These would normally be connected as shown in figure 5.6:

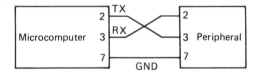

Figure 5.6

For communication in both directions, i.e. full duplex, the two handshaking control lines – 'request to send' (pin 4) and 'clear to send' (pin 5) – are also required.

The standard applies to data transmission interchange usually at rates between 110 and 9600 baud. A logic '1' is represented by a voltage in the range of −3 to −15 V and a logic '0' by a range of +3 to +15 V. This large differentiation between '1' and '0' ensures good immunity against electrical noise. However, the voltages used are not compatible with the TTL logic semi-conductor family and interconversion chips are required within the interface.

The RS232C is limited to short communication links of about 30 m although the RS422 and RS423 standards, succeeding the RS232, have extended communication distances and increased transmission speeds. The RS423, which is compatible with the RS232, has superior driving and receiving interfaces, allowing communication over distances of up to 1500 m at 9600 baud, or 15 m at 100K baud.

It should be noted that while the voltages and signal connections for the plug are defined in the standard, the data protocol is not identified. This must be known for the devices which are to be connected and can be set accordingly by software. The requirements are:

1. baud rate
2. number of bits in the ASCII group defining the character being transmitted
3. odd, even or no parity
4. number of stop bits.

For example, the default for the BBC microcomputer when set for the RS423 serial transmission port by *FX5,2 is: 1 start bit, 8 data ASCII bits, no parity and 1 stop bit. The transmission baud rate can be set with the *FX8 operating system call and the rest can be set, as required, with the *FX156 call.

The IBM-PC defaults to serial transmission with 1 start bit, 7 data ASCII bits, even parity and one stop bit for baud rates other than 110. The format can be altered by the MS-DOS command 'MODE', to suit the peripheral's requirement, and the instruction takes the form:

mode COM1 :baud, 'o' or 'e' parity, data bits, stop bits

Parallel Communication

The RS232 serial standard for communication was developed essentially for the connection of microcomputers via a telephone link. The parallel standard emerged from the need to establish a means of interfacing a variety of instruments for data logging applications. The most common standard for the integration of automated test systems, developed by Hewlett-Packard, is referred to as the IEEE-488 interface bus and has achieved wide recognition amongst instrument manufacturers since the start of the 1980s.

The bus consists of 24 lines, accommodated within standard stacked type connectors. The 8 bi-directional data lines carry information as 7-bit ASCII codes between the microcomputer (controller) and an instrument (listener) on the bus. The roles may be reversed when data is being logged. To process the information on the data bus, up to eight control and status signals are available.

The bus is designed to interface with up to 15 instruments, within a localised area, involving a total cable length of not more than 20 m. Each instrument is uniquely numbered within the range of 0–30 and the overall activity is controlled by one of the devices, termed the Controller. This is usually the microcomputer with an appropriate interface. Each device number is switch selectable within the instrument. Other functional aspects of the devices on the bus are that they must be capable of acting as a 'Listener' or a 'Talker'. A 'Listener' is a device which can receive data over the bus and a 'Talker' is one capable of transmitting data. There may be several Listeners active on the bus at any one time, but there can only be one Talker. Most devices, including the microcomputer Controller, can act as either Listeners or Talkers.

When setting up an instrument to measure some physical variable, codes devised by the instrument manufacturer are sent on the bus, in ASCII format, as a data string to the numbered device. In the case of a multichannel DVM, this could take the form of the channel number to be monitored, voltage range to be selected and a terminating character. An example of the corresponding string to be put on the bus is:

C9R2T

which denotes channel 9, range number 2, say 0–10 V, and 'T' as the string terminating character recognised by the instrument.

Manufacturers of add-on cards, to give IEEE-488 facilities with microcomputers, usually supply software for initialising the bus, setting it up for transmitting data from controller to instrument, and returning data from instrument to controller.

The measured quantity is also sent to the computer in the form of an ASCII string from which the actual numerical value can be extracted.

One of the most important management control lines is the service request (SRQ). This is a type of interrupt line that is activated low by a device residing on the bus and needing service from the controller.

Thus, a typical software sequence for implementing the control of an instrument on the IEEE-488 bus for data acquisition is:

1. Initialise bus and set instrument as a Listener.
2. Put control string on the bus to set up the instrument as required.
3. Check for SRQ line to go low indicating that data can be read.
4. Set instrument as a Talker.
5. Read returned string and convert into a numerical value.

When operating in high level BASIC, high data collection rates are not possible. However, since most instrument manufacturers offer the standard as an option, it provides an intelligently controlled flexible arrangement for test and measuring instruments.

REFERENCES

Bannister, B. R. and Whitehead, M. D., *Interfacing the BBC Microcomputer*, Macmillan Microcomputer Books, 1985.

Beebug Workshop, The 6522 timers, *Beebug*, March 1986.

Burr, B., *The Handbook of Personal Computer Instrumentation*, Burr Brown Corporation, 1988.

Bray, A. C., Dickens, A. C. and Holmes, M. A., *Advanced User Guide for the BBC Microcomputer*, Cambridge Microcomputer Centre, 1983.

De Jong, M. L., *Programming and Interfacing the 6502*, Sams, 1980.

IBM, *DOS 3.30 Reference Manual*, IBM Corp. and Microsoft Inc, 1987.

Ismail, A. R. and Rooney, V. M., *Microprocessor Hardware and Software Concepts*, Collier Macmillan, 1987.

Meadows, R. and Parsons, A. J., *Microprocessors: Essentials, Components and Systems*, Pitman, 1985.

Opie, C., *Interfacing the BBC Micro*, McGraw-Hill, 1984.

Chapter 6
Data Acquisition and Control

6.1 I/O COMMUNICATION

The input of information, such as a digital representation of a physical variable, or the output of a control signal to switch a device on or off, is accomplished through the computer ports. If the ports are 'memory mapped', i.e. each port occupying one specific byte of memory, then access is achieved by using the BASIC keywords PEEK and POKE. The contents of any byte in the computer memory can be read using the PEEK command and, similarly, the contents of any address can be altered by using POKE. While PEEK is reasonably but not always harmless, POKE must be used with some care since it is possible to access areas of memory which are used by the computer's operating system. In so doing, one might inadvertently cause the operating system to crash. If the reader is unfamiliar with memory mapped systems, then reference should be made to section 5.5 before proceeding here.

The port addresses are given in the memory map for the particular machine and can be referred to in the *User's Guide*. If, for example, the machine has two 8-bit parallel ports, one input located at address 2305 decimal and one output at address 2304 decimal, then

LET X = PEEK(2305)

will read the current 8-bit binary number at the input port and assign the appropriate numerical value to the variable called X. The LET command is optional.

The command:

POKE 2304,K

where K is some integer value between 0 and 255

will cause the current value of K to be directed to the output port. At the output port, the numerical value of K will appear in 8-bit binary form.

In the BBC microcomputer, PEEK and POKE are not available as standard keywords but have equivalents using the '?' symbol. For example:

LET X = ?&3000   equivalent to PEEK the value at the hex address of &3000 and assign the value to the variable called X
?&3020 = 240   equivalent to POKE the decimal value of 240 to the hex address of &3020

In some machines, the ports are 'port addressed', in which case there is a specifically reserved number which the computer will recognise as the port channel on a dedicated I/O bus. Access in these cases is achieved by using the keywords IN, or INP, and OUT in either BASIC or assembly language. If the input and output ports are denoted by 15 and 16 respectively, then

LET X = INP(15)

in essence, will perform the same function as PEEK in the memory mapped port. Similarly

OUT 16,K

will perform basically, the same function as POKE.

With a Versatile Interface Adaptor, such as the 6522 VIA used in 6502 micro-processor systems, the bi-directional parallel ports can be software set for either input or output. This is done using a 'Data Direction Register', i.e. DDR, or 'Control Register' which are either memory mapped or directly addressed.

In a memory mapped system, a logic '0' placed in any bit of the DDR will set the corresponding bit of the port for input. A logic '1', on the other hand, will set the corresponding bit of the port for output.

The memory mapped port locations for the 6502 based BBC microcomputer, with a 6522 VIA are:

User Port   DDR at 65122 decimal (FE62 hex)
                   Port address at 65120 decimal (FE60 hex)

Parallel   Configured as output and addressed at
Printer Port   65121 decimal (FE61 hex)

The following program, in BASIC, will set all bits of the BBC user port for input:

```
10 REM Set DDR for input
20 ?&FE62 = 0
30 REM Read current port state and print to screen
40 REM Equivalent to PEEK
50 X = ?&FE60
60 PRINT X
```

To set all bits of the port for output:

```
10 REM Set DDR for output
20 ?&FE62 = &FF
30 REM Send out a known integer value, K
40 REM Equivalent to POKE
50 ?&FE60 = K
```

Note that X and K are both integer variables in the range 0 to 255.

To set the four least significant bits (4LSBs) for input, and the four most significant bits (4MSBs) for output:

```
10 REM Set DDR for 4LSBs as input and 4MSBs as output
20 ?&FE62 = &F0
30 REM To read the 4LSBs while masking out the 4MSBs
40 X = ?&FE60 AND 15
50 REM To send out an equivalent bit sequence to output
60 K = X*16
70 REM The above is equivalent to shifting the X byte
80 REM by 4 bits to the left
90 ?&FE60 = K
```

In line 40 of the program the keyword AND is being used as a 'bit-by-bit' operator, or to use the common jargon, a 'bitwise' operator, see section 4.4.

In the example given the input is transferred to the output but must first of all be multiplied by 16 to ensure that the same binary sequence will appear on the output bits. With the X variable assigned the value of 13 the K variable becomes equal to 208, i.e. 1101 0000 binary. In this manner, any binary sequence which appears on the four least significant bits can be reproduced on the four most significant bits through multiplication by 16. This in effect is a shift left on the byte by four bits, see also section 4.1.

The equivalent assembly language routines for the three examples given in BBC BASIC are:

(i) To read the user port set as input.

```
90  [
100 LDA#0        \ Set up the DDR for input
110 STA &FE62
120 LDA &FE60 \ Equivalent to PEEK, and send the value
130 STA &FE61 \ to the parallel printer port
140 RTS
150 ]
```

(ii) To output data at the user port.

```
90  [
100 LDA#&FF     \ Set up the DDR for output
110 STA &FE62
```

```
120 LDA#&F0   \Assign some value to the accumulator and then
130 STA &FE60 \pass it to the output port
140 RTS
150 ]
```

In the above program, the value loaded into the accumulator is 240 decimal. Line 130 stores this value at the hex address of FE60, which is the user port. The equivalent binary sequence, 1111 0000, will therefore be written to the port.

(iii) To split the port to 4LSBs as input, 4MSBs as output.

```
90   [
100 LDA#&F0   \Set DDR for 4LSBs as input, 4MSBs as output
110 STA &FE62
120 LDA &FE60 \Equivalent to PEEK the input at the port,
130 AND #&0F  \with 4MSBs masked
140 STA &FE61 \Write value to the parallel printer port
150 ASL A ⎫
160 ASL A ⎬  \Shift the accumulator 1 bit to the left
170 ASL A ⎪
180 ASL A ⎭
190 STA &FE60 \POKE value to the output port
200 RTS
210 ]
```

Lines 150 to 180 have the same effect that multiplying the input by 16 had in the BASIC program.

These above assembly language programs for a BBC microcomputer are all incomplete since the assembler has been given no indication of where to store the assembled machine code. The means of allocating memory, setting the program counter and calling a machine code subroutine from a BASIC program are detailed in section 4.7.

6.2 DIGITAL INTERFACING

When a bit at the output port of a computer is set to logic 1, it is said to go 'high' and the voltage at the bit is approximately 5 V. The current available however is minimal, of the order of 1 mA, and no load can be connected directly to the port, however small. There is an imminent danger, in fact, of damaging the computer if a load is connected to the port and a suitable interface must therefore be provided to enable computer control of the power switching of loads. Such a device is a power transistor which operates off the output port control signal. The power transistor can be used to switch a conventional electro-mechanical relay, which in turn can switch in a load. The load may be some or other 240 volt a.c. mains device or perhaps a d.c.-operated solenoid-controlled pneumatic or hydraulic valve.

A particularly suitable integrated circuit for this interface is the 'Darlington driver', ULN2001A, which incorporates a number of Darlington transistor pairs and can sink up to 500 mA at 50 V. These ICs are available from most electronic component manufacturers. The Darlington driver, such as RS 307-092, incorporates seven separate stages, each of which is diode protected for the switching of inductive loads. The chip interfaces directly with the TTL and CMOS logic families as used in microcomputer systems.

The wiring diagram for one stage in a Darlington driver is illustrated in figure 6.1.

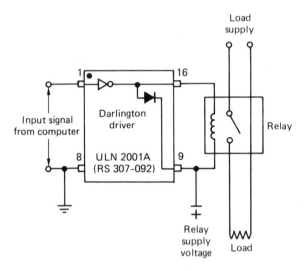

Figure 6.1 *Darlington driver*

To provide total isolation from high voltages, signals can be coupled through an 'opto-isolator', ISQ74, between the output port of the computer and the power control device. The opto-isolator negates the need for 'hardwired' electrical connections and the signals are transmitted by means of a beam of infra-red radiation, emanating from an emitting device and 'sensed' at a detector. The beam, in fact, simply switches in a transistor at the output pin and a number of stages may be incorporated within one IC. For example, RS 307-064 has four stages.

It should be noted that an opto-isolator is by no means an essential element in an interface and that it is purely a protective device to combat spurious noise signals which can corrupt the digital logic values being transmitted on the buses. Figure 6.2 gives a schematic representation of a single stage in a typical opto-isolator.

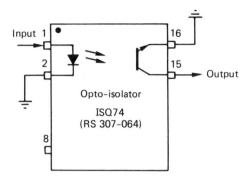

Figure 6.2 *Opto-isolator*

When a microcomputer based on TTL circuitry is powered up, the state of the bits of the user port 'float high'. In other words, each bit of the user port becomes set to a logic 1. The printer port however defaults to output, with all bits of the printer port set low. To counter the peculiarity of the user port floating high, an inverter chip can be included between the port and all other external hardware. The inverter performs the function of inverting all logic signals of 1 to 0, or vice-versa. Thus when the computer is switched on, all 8 bits of the user port will float high, with a logic 1. The inverter, however, will ensure that all 8 bits on its output pins will have a logic 0. This is a safety feature which prevents the sending of a logic 1 control signal and thereby inadvertently switching on some power device. Following power-up of the computer, a logic 0 must then be sent to the relevant bits of the output port, to become a logic 1 after inversion and to operate the control function, i.e. an output of '0' signifies 'on' and a '1' signifies 'off'. A common inverter chip used is one from the 74TTL logic family. The 7404 hex inverter (such as RS 305-529) has six stages and is packaged as a 14 pin dual-in-line (DIL), semi-conductor IC, see figure 6.3.

The basic elements comprising the computer output port interface and providing safety, isolation and amplification are:
  (i) Hex inverter to eliminate inadvertent switching of power.
 (ii) Opto-isolator to prevent corruption of logic signals on the data bus and to provide protection for the computer internals from stray voltage.
(iii) Darlington driver to amplify the port control signal and enable it to switch in an electro-mechanical relay, which can in turn switch in a load.

In high frequency switching applications, electro-mechanical relays are unsuitable and semi-conductor devices such as silicon controlled rectifiers (SCRs), alternatively called thyristors or triacs, may be appropriate. Also available are the solid state relays (such as RS 346-671), which operate directly from a TTL signal typical of that available at a computer user port. The latter are particularly suited to a.c. loads and since their switching is implemented at the so-called

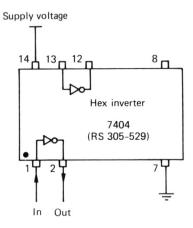

Supply voltage

Figure 6.3 *7404 hex inverter*

'cross-over point' when no current is flowing, then no surge is experienced by the computer.

All of the elements considered are readily available from many electronic components suppliers. Data sheets are also normally supplied on request, giving the component specification, details of typical applications and comprehensive wiring diagrams.

A complete digital interface for a computer output port, suitable for load switching, is illustrated in figure 6.4.

6.3 ANALOGUE INTERFACING

Section 6.2 dealt with the information transfer from the computer to external, or peripheral, devices. The transfer of information in the other direction, from peripheral devices to the computer, is the essence of data acquisition and condition monitoring. In most practical applications it is the monitoring and acquisition of the data which is the necessary precursor to the subsequent control functions which might be implemented.

In the experimental context the physical variables are represented ultimately as voltages from the measuring transducers. Using the methods outlined in chapter 2, these voltages would normally be arranged to vary between 0 volts and some suitable reference voltage. This would represent the full range of the physical variable and constitutes the basic analogue signal which requires to be 'digitised' before manipulation and processing by the microcomputer. The task of signal conversion is performed by the analogue-to-digital converter (ADC or A/D converter), which samples the analogue input, performs the conversion and

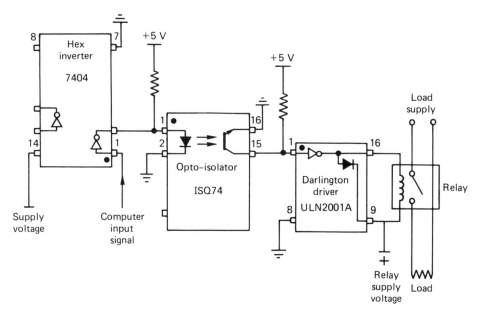

Figure 6.4 Output port digital interface

outputs a digitally encoded number which is directly proportional to the magnitude of the analogue input. The basic process is illustrated in figure 6.5.

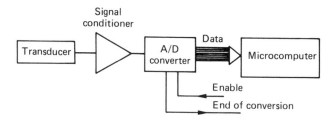

Figure 6.5 Basic analogue interface

There are a variety of A/D converters available on the market and the more common are:

(i) Staircase and Comparator

In this type of converter a voltage is generated which is gradually increased in small steps, i.e. a 'staircase' waveform. The magnitude of this voltage is compared with that of the input voltage at each successive step. When the generated voltage is approximately equal to the input, the process is stopped and a binary count of the number of steps taken for the staircase waveform to reach the level of the input voltage is then made. The binary count from zero is presented at the output as the digitally encoded representation of the analogue signal. The process is depicted in figure 6.6.

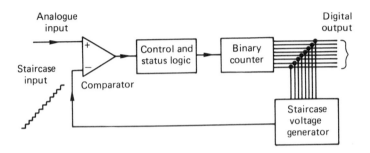

Figure 6.6   Staircase and comparator A/D converter

The staircase waveform is in itself generated by a digital-to-analogue (D/A) converter which performs the reverse function of the A/D converter. The operating principle of the D/A converter can be illustrated as a simple example of Ohm's law.

Considering a 4-bit system with bits b_0 to b_3. If the current I_0 represents b_0, $2I_0$ represents b_1, $4I_0$ represents b_2, and $8I_0$ represents b_3, then for a constant supply voltage V the resistance R must be halved each time a digit moves one bit towards the MSB. That is

$$I_0 = V/R \qquad = I_0$$
$$I_1 = V/(R/2) \quad = 2I_0$$
$$I_2 = V/(R/4) \quad = 4I_0$$
$$I_3 = V/(R/8) \quad = 8I_0$$

The required circuitry is shown in figure 6.7.

In the diagram the digital input is 1011 with the 100k, 50k and the 12.5k resistors connected in parallel. The 25k resistor is not switched into the circuit.

The total effective resistance is given by:

$$R_{\mathrm{T}} = \cfrac{1}{\cfrac{1}{100} + \cfrac{1}{50} + \cfrac{1}{12.5}} = \frac{100}{1 + 2 + 8} = \frac{100}{11}$$

Hence $I_{\mathrm{REF}} = V/R_{\mathrm{T}} = \dfrac{10 \times 11}{100} = 1.1$ mA

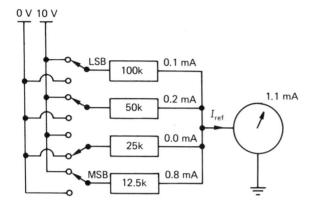

Figure 6.7 *4-bit D/A converter*

It is apparent that the currents are additive such that any digital input in the range 0 to 15 will result in an output in the range 0 to 1.5 mA, in direct correspondence with the digital input. This provides a staircase current function which can be converted to a staircase voltage by the addition of suitable operational amplifier-based circuitry.

The staircase and comparator type of A/D converters have relatively slow conversion rates, of the order of 20 ms, but they are essentially immune to electronic noise and are comparatively inexpensive.

(ii) Integrating Type (or Dual Slope)

The major elements comprising a dual slope A/D converter are illustrated in figure 6.8.

At the start of conversion a voltage-to-current converter is switched to the integrator, causing it to ramp up a slope which is proportional to V_{in}. This occurs over a fixed period of time at the end of which the input is switched over to the reference current source. At the instant of switching the integrator output voltage is proportional to V_{in}, a counter is enabled and counting begins at a rate set by the internal clock. In the meantime, the reference current causes the integrator to ramp down at a slope which is proportional to V_{ref}, i.e. a constant slope. When the integrator output again reaches ground, the comparator switches the counter off and the counter then contains a digitally encoded value proportional to V_{in}. Figure 6.9 shows the voltage variation at the integrator output.

From figure 6.9 it can be seen that there are two similar triangles such that:

$$\frac{V_{ref}}{T_m - T_f} = \frac{V_{in}}{T_v - T_f}$$

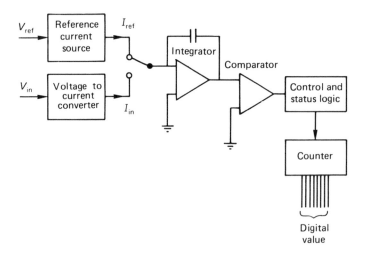

Figure 6.8   Dual slope A/D converter

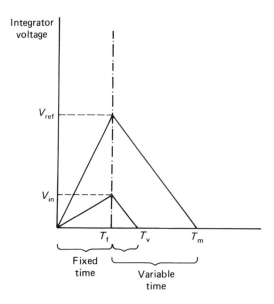

Figure 6.9   Integrator voltage variation

$$\therefore \; V_{in} = \frac{T_v - T_f}{T_m - T_f} \, V_{ref}$$

T_v is directly proportional to the counter output and with T_m, T_v and V_{ref} all known, the input voltage, V_{in}, is determined by proportion.

The integrating type of A/D converters have similar operating characteristics and conversion times to that of the staircase and comparator types. For faster A/D conversion, the 'successive approximation' or 'counter' type are generally employed.

(iii) Successive Approximation Type

In this A/D converter, the input signal is compared with a number of standard reference voltages, generated from a D/A converter, until the combination of standard voltages required to make up the input value has been determined. The main components of the converter are a clock, a counter, a comparator and a D/A converter.

When an analogue signal is input to the converter, the counter starts a count and passes a digital value to the D/A converter. The D/A converter generates a voltage to represent the most significant bit and the comparator assesses this against the analogue input. If the analogue signal is greater than the voltage from the D/A converter then the logic 1 in the MSB is retained. If the analogue signal is smaller then a logic 0 is assigned to the MSB. This process is then repeated on the next most significant bit and so on for all the other bits down to the LSB. The conversion time for these types of converters may be of the order of 10–25 μs, but this will depend upon the hardware design. Figure 6.10 outlines the essential features of a successive approximation A/D converter.

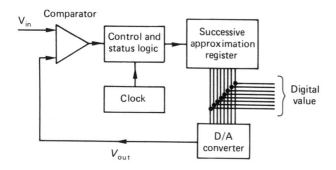

Figure 6.10   Successive approximation A/D converter

(iv) Parallel Conversion Type

The parallel type A/D converter has by far the fastest conversion time, at about 1 μs, but it is also the most expensive. With parallel conversion, the analogue input is fed simultaneously to a number of comparator circuits, each having a different reference voltage. The resulting comparator outputs are fed to a logical coding network which generates the appropriate digital values to represent the state of the comparator outputs.

Regardless of the type of A/D converter used, the pin functions on the IC are basically similar and generally comprise the power supply, the data bits, the start conversion pin, ($\overline{SC}$ or $\overline{CONVERT}$) and the end of conversion pin, ($\overline{EOC}$ or $\overline{STATUS}$). The 'overscore' signifies that the pin is active low.

The conversion is software initiated by sending a 'pulse' (logic 0, followed by logic 1) to the $\overline{CONVERT}$ pin. On the negative edge of this pulse the counter in the successive approximation A/D converter is set to zero, and on the positive edge the counter starts incrementing. At the start of conversion the $\overline{STATUS}$ pin goes from low to high and when it again goes low, the conversion is complete.

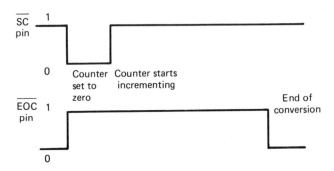

Figure 6.11 '*Start conversion' and 'end of conversion' pin signals*

The end of conversion may be readily detected using suitable software. As an alternative it is possible to include a time delay following the start conversion pulse to allow conversion to complete before reading the value at the input port. The length of the delay can generally be found by trial and error.

In choosing the appropriate A/D converter for a particular application the four main features to be considered are:

1. Conversion Time

The conversion time is a measure of the operating speed of the converter and is the time taken for the complete translation of an analogue signal to digital form. The conversion time in many of the staircase and comparator and the integrating types of A/D converter may be dependent on the level of the analogue input signal. Faster conversion times are obtained with low level inputs because of the manner in which the conversion is completed. Successive approximation and parallel conversion types of A/D converter have a fixed conversion time. This is because the exact same conversion process is performed regardless of the analogue input level.

2. Resolution

The resolution of an A/D converter is the number employed to represent the digital output in binary form. The resolution, for example, of an 8-bit A/D converter is limited to one part in 255 of the maximum voltage corresponding to the full scale setting. An improvement in resolution can be obtained with a 12-bit converter, with one part in 4095. Table 6.1 summarises the relation between the number of bits and the resolution.

Table 6.1

n-bits	$2^n - 1$	Resolution (%)
8	255	0.4
10	1023	0.1
12	4095	0.025
16	65535	0.0015

3. Accuracy

The accuracy is related to linearity defects, zero error and calibration deficiencies in the electronics of the converter and should not be confused with the resolution.

4. Cost

Cost will depend on the quality required in the three areas previously described and on the means of conversion employed. The cost is closely associated with the speed of the conversion and with the resolution and accuracy. Cost generally increases with increases in all, or either, of the three other variables considered.

Table 6.2 gives a comparison between various A/D converters.

Table 6.2

Type no.	RS number	n-bits	Conversion rate	Linearity	Type
TL507C	304-150	7	1 kHz	± 1.0 LSB	Integ. single slope
CA3318E	636-665	8	20 MHz	± 1.0 LSB	Parallel conv.
ZN439E-7	633-694	8	0.2 MHz	± 1.0 LSB	Succ. approx.
ZN427E-8	309-464	8	0.1 MHz	± 0.5 LSB	Succ. approx.
8703CJ	308-045	8	0.8 kHz	± 0.5 LSB	Integ. dual slope
AD7572JN-12	637-753	12	0.2 MHz	± 1.0 LSB	Succ. approx.
AD7578KN	637-775	12	10.0 kHz	± 1.0 LSB	Succ. approx.
ADC-302	655-302	8	50.0 MHz	± 0.5 LSB	Parallel conv.

6.4 MULTIPLEXING

In applications where a number of transducers are to be sampled by a microcomputer, a multiplexer (MUX) can be used to switch in various channels as and
when required to a single A/D converter. The switching is software controlled
from the microcomputer and figure 6.12 illustrates the basic principle.

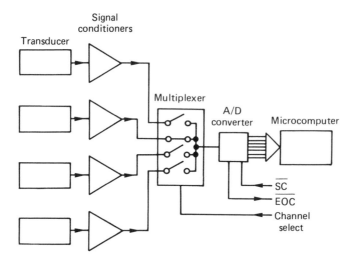

Figure 6.12 *Multiplexer for multiple inputs*

The multiplexer and A/D converter often form an integral part of the microcomputer system. In some cases, even the signal conditioning can be software
controlled, with all the necessary hardware mounted on a 'card' and plugged
directly into the computer's bus system.

Multiplexers, or analogue switches, are available with various numbers of
input channels, e.g. DG508ACJ, such as RS 309-571, has eight analogue input
lines. Channel selection on the RS chip is implemented through three control
lines operating on logic level signals.

A typical application, illustrated in figure 6.13, is an energy management
system for a motor vehicle. The microcomputer-based controller contains the
manufacturer's algorithm to implement the decision as to when the distributor
should issue the spark depending on the state of the various physical variables as
measured.

Minimum cost conditions usually dictate whether multiplexing will be implemented or not, but the reduced cost must be balanced against an inevitable
reduction in sampling rate. Figure 6.14 shows three possible arrangements of
signal conditioning, multiplexing and conversion for analogue interfaces.

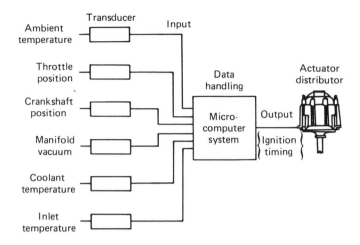

Figure 6.13 Energy management system for a motor vehicle

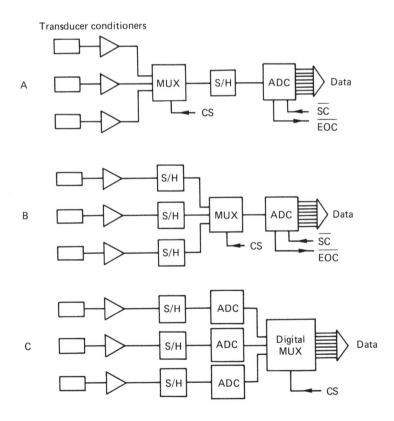

Figure 6.14 Multi-input systems

The sample and hold (S/H) function is provided to capture the amplitude of the input analogue signal at a particular period in time. The signal may still be varying with time. Many A/D converters have the sample and hold function in-built.

System A is the most common, while B and C can provide for virtually simultaneous sampling. System C gives the most representative snapshot at a particular period in time, but it is also the most costly.

6.5 DIGITAL SAMPLING

An important aspect relating to any digital data-acquisition system is the sampling rate. The problem is really double-edged in that while the sampling rate must be fast enough to capture an authentic representation of the input signal, it cannot be as fast, or faster, than the A/D conversion rate.

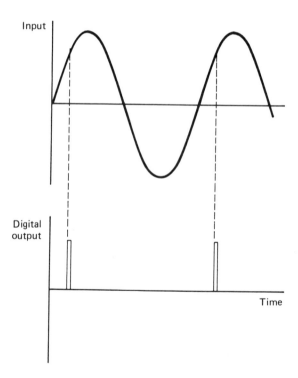

Figure 6.15   Digital sampling at input frequency

Consider a sinusoidal input of say 1000 Hz. If the sampling rate has the exact same frequency, then the digital approximation will be manifest as a constant d.c. level as illustrated in figure 6.15.

The apparent d.c. level will depend upon the phase relationship between the input signal and the sample time.

If the sampling frequency was to be increased to 1001 Hz, then the sampled point would progress along the input signal such that over a period of one second the sampled points would describe a sine wave with the correct amplitude, but with a frequency of 1 Hz. This type of error is known as aliasing. To avoid aliasing errors the sampling frequency should, according to Shannon's sampling theory, be at least twice that of the most dominant frequency in the input signal. A more practical criterion would be a factor of five to ten times the most dominant frequency. If the frequency content of the input signal is not known prior to testing, a lowpass filter should be employed to attenuate all of the input signal with frequencies above half of the sampling frequency, see section 2.5. In this way, if it is appropriate as far as the measurement of the physical variable is concerned, aliasing errors can be effectively eradicated.

The integrating type of A/D converter has a slow conversion time and this limits the application of the converter to a maximum sampling rate, typically about 1000 Hz. Using a factor of five, to avoid aliasing errors, the signal frequency which can be captured is therefore of the order of about 200 Hz. Many other integrating type of A/D converters are considerably slower than this and for faster data-acquisition applications the successive approximation type of A/D converter must be used. Modern versions of the successive approximation type of A/D converter can operate down to the order of one micro-second conversion time, i.e. a 1 MHz conversion speed. It is never possible to achieve the same sampling speed, however, since there must always be some 'overhead' time allowance for storing the acquired data and updating the storage byte locations. The typical conversion process is:

Initiate the $\rightarrow$ wait for $\rightarrow$ digital output $\rightarrow$ store data and
A/D converter   conversion   from converter   update counter

The last function is a characteristic of the microcomputer and depends on the type of processor and how the wait for conversion time may be utilised.

To maximise the sampling rate, assembly language data-acquisition routines must be used in conjunction with an external successive approximation type A/D converter. The forms of the external A/D converter are varied. They may range from a custom built circuit to interface with a user port, to a complete add-on peripheral system, or an expansion card.

A custom built A/D interface for the BBC user port is featured in section 10.2. A/D expansion cards for the IBM-PC are covered in section 8.4 and add-on peripheral systems are covered in section 6.7.

6.6 DIGITAL CONTROL

The impact of the microprocessor has made the control of systems and processes by microcomputer a normal feature. The programmable digital controller may be defined as:

"A digitally operating electronic apparatus which uses a programmable memory for the internal storage of instructions for implementing specific functions to control various types of machines or processes. The functions could include logic, sequencing, timing, counting and arithmetic and the communication with the computer is through a variety of digital or analogue modules."

A microcomputer system neatly fills the above specification.

Academic control is centred round the mathematical modelling of systems and processes. A detailed knowledge of the system components is therefore required to formulate the differential equations which will describe the system behaviour for various external input conditions. These qualitative models give rise to transfer function descriptors and step, or frequency, response methods are normally used to assess the system performance. The major problem in the classical approach is that it can be very difficult, sometimes impossible, to derive an adequate transfer function to describe the system. For this reason recent developments have leaned towards computer based control strategies which can be made to work with real systems.

The basic elements of the computer control system include:
 (i) the computer, with associated processor
 (ii) memory and input/output facilities
(iii) transducers for monitoring the system condition
(iv) actuators for valves, motors and other mechanical hardware.

In addition, software for the specified control of the system must be written to interact with the chosen hardware. For this to be successful, a thorough understanding of the system operation becomes essential.

A number of microcomputer control applications are included to illustrate the basic principles. Most of the examples in this section use the BBC microcomputer as the vehicle for illustration. The same examples are repeated for comparison in chapter 8, using the IBM-PC as the host computer fitted with an appropriate expansion card.

1. Sequential Control

This type of control is used over a wide spectrum of applications and may be either event- or time-based. Event-based sequence control depends on one action taking place only when the previous action has been completed. This provides interlocking in the sequence which will cause the cycle to halt should any particular action not be completed. In time-based sequence control, the events are ordered relative to time.

Example

An illustrative example of this strategy is the sequencing of a single pneumatic cylinder, controlled by solenoid-operated valves as shown in figure 6.16. The end of stroke positions are sensed by the microswitches 1 and 2 and solenoid A must be held on to supply air to the system. When solenoid B is energised, the piston will move to the left. Alternatively, when solenoid C is energised, the piston will move to the right.

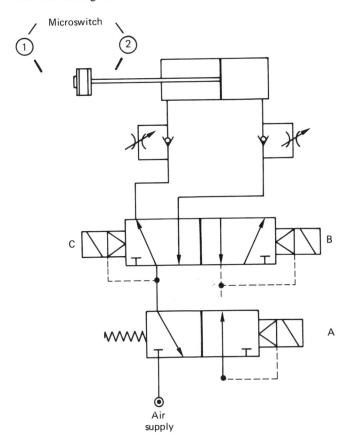

Figure 6.16 Event-based sequence control

This system requires two input and three output control lines and these could be connected to the computer ports as indicated below:

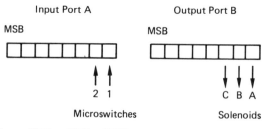

(address = 2305   : DDR = 2307)   (address = 2304   : DDR = 2306)

The microswitches can be connected such that a logic 0 signifies that the switch has been activated and the output port can be connected to a power scaling interface which activates the solenoid valves when a logic 0 is output.

A flow diagram and the corresponding control program in BASIC, for the continuous cycling of the cylinder, is given below. It is assumed that the cylinder starts from the retracted position.

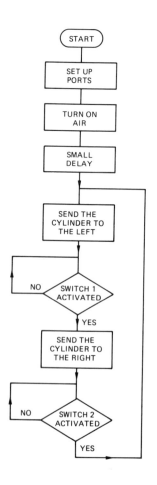

```
10 REM PORT A INPUT
20 POKE 2307,0
30 REM PORT B OUTPUT
40 POKE 2306,255
50 REM SWITCH AIR ON
60 POKE 2304,254
70 FOR K=1 TO 200:NEXT K
80 REM SEND CYLINDER TO THE LEFT
90 POKE 2304,252
100 REM WAIT UNTIL END OF STROKE
110 IF (PEEK(2305) AND 1)=1 GOTO 110
120 SEND CYLINDER TO THE RIGHT
130 POKE 2304,250
140 REM WAIT UNTIL END OF STROKE
150 IF (PEEK(2305) AND 2)=2 GOTO 150
160 GOTO 90
```

If ten cycles are to be completed then the following lines should be added:

75 FOR I = 1 TO 10
155 NEXT I

Line 160 should be removed

This exercise could be carried out using the user port on the BBC micro-computer as follows:
Set the 4 LSBs as output, the 4 MSBs as input and connect the input/output port as shown below:

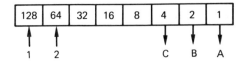

address = &FE60   : DDR = &FE62

Program for continuous cycling:

```
10 ?&FE62=15
20 ?&FE60=254
30 FOR K=1 TO 1000:NEXT K
40 ?&FE60=252
50 IF (?&FE60 AND 128)=128 THEN 50
60 ?&FE60=250
70 IF (?&FE60 AND 64)=64 THEN 70
80 GOTO 40
```

Alternatively, in 6502 assembly language and omitting the delay routine, the program listing is:

```
LDA#&F          \ set port for 4 LSBs output
STA &FE62       \ and 4 MSBs input
LDA#&FE         \ switch on the air
STA &FE60       \
.more
LDA#&FC         \ move cylinder to the left
STA &FE60
.SW1
LDA &FE60       \ wait until switch 1 activated
AND#&80
BNE SW1
LDA#&FA         \ move cylinder to the right
STA &FE60
.SW2
LDA &FE60       \ wait until switch 2 activated
AND#&40
BNE SW2
JMP more        \ repeat the sequence
```

The movement in pneumatic systems is relatively slow so there is no real advantage to be gained in using low level programming. However, if the software is to be run from EPROM, it would first of all have to be written in assembly language and assembled into machine code prior to blowing the EPROM.

2. ON/OFF Control

This is a simple control strategy where the controller action alternates between either fully on or completely off as the controlled variable's value changes respectively from below, or above the specified desired condition.

Example

A typical exercise on the ON/OFF control of a power switching process is the temperature control of a quantity of water. The power devices are a 240 V a.c. heater and a 24 V d.c.-operated cooling fan as indicated in figure 6.17.

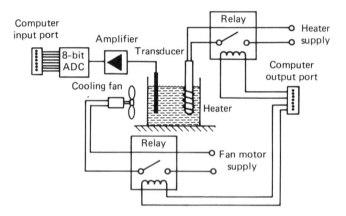

Figure 6.17   ON/OFF control system

The temperature sensor is connected to an amplifier which is set to give an input signal of 1.8 V to the A/D converter, corresponding to a temperature of 100°C. Zero volts corresponds to 0°C and the relationship between the digital output value and the temperature is linear.

Using computer control, the temperature of the water is to be maintained at 40°C ± 0.5°C and the following ON/OFF strategy can be applied:

$$T < 39.5°C, \text{ then heater on and fan off}$$
$$39.5°C \leqslant T \leqslant 40.5°C, \text{ then both heater and fan off}$$
$$T > 40.5°C, \text{ then heater off and fan on}$$

The two switching control signals for the heater and the fan are taken from the output port LSBs, these being P0 and P1 respectively. Additionally, a power switching interface is required, see section 6.2.

The interface would incorporate an inverter, opto-isolator, Darlington driver and possibly an electro-mechanical relay. Alternatively, the 240 V a.c. power may be switched by a solid-state relay, operating off the port output control signal. In either case, a logic 0 is taken to signify ON.

The software is easy to write once a flow diagram has been drawn up to conform with the specified control strategy.

Using the BBC microcomputer and channel 1 of the analogue input port, see section 7.1, to monitor the water temperature:

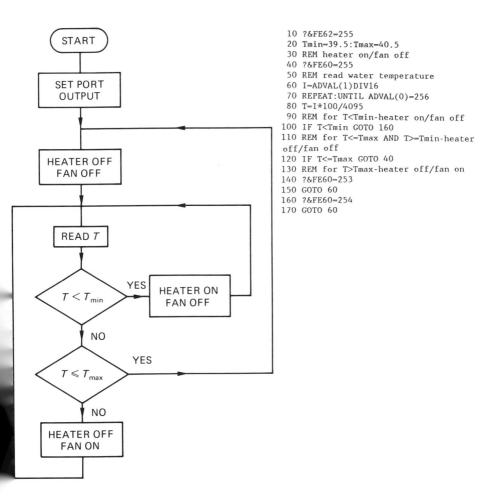

```
10 ?&FE62=255
20 Tmin=39.5:Tmax=40.5
30 REM heater on/fan off
40 ?&FE60=255
50 REM read water temperature
60 I=ADVAL(1)DIV16
70 REPEAT:UNTIL ADVAL(0)=256
80 T=I*100/4095
90 REM for T<Tmin-heater on/fan off
100 IF T<Tmin GOTO 160
110 REM for T<=Tmax AND T>=Tmin-heater
off/fan off
120 IF T<=Tmax GOTO 40
130 REM for T>Tmax-heater off/fan on
140 ?&FE60=253
150 GOTO 60
160 ?&FE60=254
170 GOTO 60
```

The ADVAL and other related commands in BBC BASIC are explained in more detail in chapter 7.

A useful extension to the above exercise is to give a continuous plot of the temperature against time on a display monitor.

3. Closed Loop Feedback Control

This involves a control strategy based on the error signal which is the difference between the desired process variable value, termed the set point (SP), and the currently measured process variable value (PV), that is

$E = \text{SP} - \text{PV}$

When a digital computer is used as the controller then an A/D converter is interfaced to the input to convert the analogue signal from the measuring transducer. A D/A converter is interfaced to the output to provide the variable analogue signal to control the process.

The single negative feedback closed loop control of a process can be illustrated in block diagram form as follows:

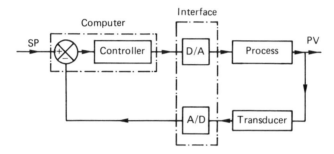

Figure 6.18 *Negative feedback, closed loop control*

The most common control strategy used is the so-called three-term controller with the controller settings evaluated from either the open loop or the closed loop response of the actual system.

The three constituent parts of the controller action, based on the evaluated error, are as follows:

(i) Proportional Action

controller output $= K \times E$

where K is termed the controller gain.

For proportional control only there must be an error signal in order to produce an output control action. Theoretically the gain must be infinite if the set point is to be reached.

In practice, companies which supply controllers tend to favour the term proportional band (PB) in preference to controller gain. Proportional band is related to gain by:

$$\text{PB\%} = \frac{100}{K}$$

Consider the control of temperature by proportional action. The temperature range is from 0°C to 100°C and a proportional band of 2 per cent is to be used. The set point, i.e. the desired temperature, is to be say 60°C. Figure 6.19 shows the controller output.

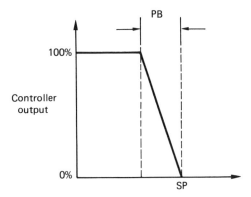

Figure 6.19 Proportional band controller output

The proportional band is 2 per cent of 100°C, i.e. 2°C. If the temperature is less than (SP − PB), i.e. (60 − 2) = 58°C, then 100 per cent of the available power would be supplied to the heating device. As soon as the temperature reaches 58°C, the amount of power supplied to the heater would be reduced in proportion to the temperature and how close it is to the set point. At 59°C, the power supplied to the heater would be 50 per cent of the maximum available. When the temperature reaches the set point of 60°C, the heater is switched off, i.e. 0 per cent of the available power is supplied to the heater. It is clear that with a very small proportional band, the action is tending towards a simple ON/OFF strategy with either 0 per cent or 100 per cent of power supplied to the heater. A large proportional band will result in a much slower response of the system in reaching the set point.

(ii) Integral Action

The limitation of proportional control can be overcome by adding integral controller action which gives a controller output contribution proportional to the integral of the error signal.

$$\text{Controller output} = K_i \int E dt$$

where K_i is the controller integral gain usually expressed as K/T_i with:

   K = controller gain, and

   T_i = controller integral time constant in seconds.

Hence

$$\text{integral contribution to output} = K/T_i \int E \, dt$$

It can be seen that the controller output increases as long as an error exists, but as the error tends to zero the controller output tends to a steady value.

If T_i is large, the response to the integral action is slow and the error may persist for too long.

If T_i is too small, the magnitude of the integral term may cause excessive overshoot or even cause the system to become unstable resulting in the process variable value continually increasing with time.

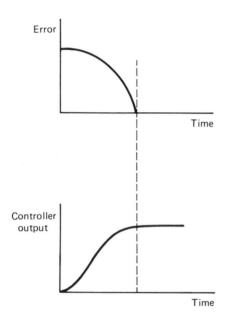

Figure 6.20   Controller response for integral action

(iii) Derivative Action

The stability of the system can be improved and any tendency to overshoot reduced by adding derivative control action which is based on the rate of change of the error.

$$\text{Controller output} = K_d \times \frac{dE}{dt}$$

where K_d is the controller derivative gain usually expressed as $K T_d$ with:

K = controller gain, and

T_d = controller derivative time constant in seconds.

The full three term (PID) control strategy is therefore:

$$\text{Controller output} = K \left[E + \frac{1}{T_i} \int E \, dt + T_d \times \frac{dE}{dt} \right]$$

With error E = (required set point SP − measured value of PV), the controller settings are: gain K, integral time T_i and derivative time T_d.

Essentially, the proportional action governs the speed of response, the integral action improves the accuracy of the steady state, and the derivative action has a stabilising effect.

A number of manufacturers of digital controllers offer a software facility for implementing three term (PID) control based on the above strategy.

The corresponding digital control algorithm based on discrete time steps of Δt is:

$$\text{output} = K [E_i + ((\Delta t / T_i) \, \Sigma \, E_i) + (T_d / \Delta t) \, (E_i - E_{i-1})]$$

where $E_i = \text{SP} - \text{PV}_i$

In deriving the digital control algorithm in the form of finite differences, errors arise because of the choice of sampling interval Δt. The character of the digital algorithm therefore differs from that of its continuous counterpart. Typical values for Δt are based on empirical rules and for control loops involving flow, pressure, level and temperature a sampling interval of 1 second is normally satisfactory. Faster acting systems, for example, the control of position or speed, use sampling intervals down to a few milliseconds.

Controller setting values of K, T_i and T_d can be obtained according to the empirical methods of Zeigler and Nichols.

Example

Consider a process which requires water to be supplied from a constant-level tank with the in-flow to the tank coming from a pump through a flow control valve. Control of the level is to be carried out using a microcomputer with A/D and D/A input and output facilities. The water level is measured by a level sensor which outputs a voltage signal and the flow from the pump is varied by a flow control valve which operates off pneumatic pressure in the range 3–15 p.s.i. This is the normal arrangement for a flow control valve and manufacturers usually incorporate a current to pressure (I/P) converter into the valve control hardware, equating 4 mA to 3 p.s.i. and 20 mA to 15 p.s.i. In order to control the valve setting from a DAC output, a voltage-to-current converter (V/I) is also required. For example, 0 V gives 4 mA and 10 V gives 20 mA. This type of hardware is normal in digital control systems where analogue measurements and analogue-based control hardware are still employed.

The system is shown diagrammatically in figure 6.21.

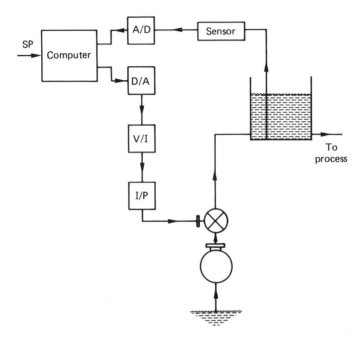

Figure 6.21   *Process control system*

Sensor: 0 V equivalent to height of 0 mm
          1.8 V equivalent to height of 1000 mm
ADC: 12-bit uni-polar, 0 to 1.8 V
DAC: 8-bit, 0 to 10 V
          10 V equivalent to valve fully open
          0 V equivalent to valve fully closed
Required level to be 500 mm (SP)

A flow diagram and program for the closed loop control of the system is given below for a control loop taken over a time interval of 1 second. On a BBC microcomputer, the parallel printer port may be used to operate the external D/A converter.

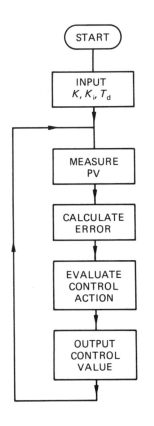

```
 10 INPUT"LEVEL REQUIRED(mm)= "SP
 20 INPUT"GAIN(K)= "K
 30 INPUT"INTEGRAL TIME(s)= "Ti
 40 INPUT"DERIVATIVE TIME(s)= "Td
 50
 60 *FX16,1
 70 Dt=1:SUM=0:EP=0
 75 IC=Dt/Ti
 76 DC=Td/Dt
 80 TIME=0
 90 PROCpid
100 REPEAT:UNTIL TIME=100
110 GOTO 80
120
130 DEF PROCpid
140 REPEAT:UNTIL ADVAL(0)=256
150 REM 1000mm HEIGHT EQUIV.TO 4095
160 PV=(ADVAL(1)DIV16)*1000/4095
170 E=INT(100*(1-(PV/SP)))
180 SUM=SUM+E
190 outputI=IC*SUM
200 outputD=DC*(E-EP)
210 output=K*(E+outputI+outputD)
220 IF output<0 THEN output=0
230 IF output>100 THEN output=100
240 REM output control value from
    printer port as an 8-bit number
250 ?&FE61=output*2.55
260 EP=E
270 ENDPROC
```

The program is illustrative of how the three term control algorithm is translated into software. Without having the physical system hardware, however, it is
difficult to visualise the effectiveness of the control action for various control
parameter inputs. Some idea of the controller action can be obtained by connect-
ing a potentiometer across the reference voltage of the onboard A/D converter
on the BBC microcomputer. This may be used as an input to channel 1 of the
analogue port. The error and 'output' value may be printed to the screen as
percentages.

For example, take SP as 500 mm say and eliminate integral and derivative
action with T_i = 1E9 and T_d = 0. Take a gain of 2 and start with a level of
0 mm equivalent to 0 V. Observe the output control action value and adjust the
potentiometer accordingly. A large number indicates high flow and fast level
increase while a small value means low flow and a level decrease, when the in-
flow is less than the out-flow.

6.7 PERIPHERAL SYSTEMS, DATA LOGGING

The term 'peripheral system' can be applied to any hardware device which can be considered as additional to the basic host microcomputer. The peripheral system may take the form of a 'data logger' which can be programmed to capture and store measurements from a range of transducers. Some battery powered, stand alone data loggers may be programmed independently. At a later date they may be coupled to a microcomputer to down-load the captured data for analysis and display. Typical of this type of data logger is the 'Squirrel meter/logger', available from Grant Instruments. Up to 19 input channels can be made available and the user must specify the number and type of inputs required. The inputs might typically include voltage, current, temperature, humidity, pulse count and rate and digital event marking. The user must also specify the memory size and input ranges required and whether the inputs are to be averaged or instantaneous.

Programming the Squirrel is quite straightforward and data transfer software is available for most of the popular microcomputers including the IBM-PC and the BBC. Communication with the microcomputer is through the RS232C serial transmission link and the Squirrel is marketed in 8-bit and 12-bit versions. The system can be expensive however, depending on the specification. The advantage of the Squirrel meter/logger lies in its capacity for remote, unsupervised monitoring applications.

Almost all other data logging hardware is less portable than the Squirrel and these systems are designed to be operated in conjunction with the host microcomputer on-site. Such systems include the INLAB/THINKLAB interface, available from 3D Digital Design and Development Ltd. This is a modular system which can be built up to suit the user's requirements and can be supplied with compatible interfaces for most of the popular microcomputers. Signal conditioning modules with a variety of amplification and multiplexing functions are available and a wide range of A/D conversion modules can be specified. Additional modules are also available including analogue output, digital I/O, stepper motor controllers, shaft encoder interfaces and frequency/time measurement interfaces, among others. Complex modular systems can be very expensive but a range of cheaper general-purpose systems is also available. A typical general-purpose INLAB system suitable for IBM-PCs, BBCs and a range of other microcomputers might include the following:

(i) 8 differential, or 16 single ended analogue input channels

(ii) 12-bit A/D converter with 8 μs conversion time

(iii) input voltage range, $\pm$ 100 mV to $\pm$ 10 V

(iv) gain control through a multi-turn potentiometer

(v) 8 digital output lines, 8 digital input lines.

The INLAB, or similar general-purpose interface system, would normally be supplied as a boxed unit, complete with all the necessary connectors and illustrative software to run with the appropriate microcomputer.

Many other similar systems for real time data-acquisition and control are marketed. Some popular systems with IBM-PC or BBC compatibility are listed below:

Table 6.3   Peripheral systems for data acquisition and control

Trade name	Manufacturer	Typical system
OASIS MADC12	Peter Nelson Design Consultancy, NORWICH NR8 5DF	16 channel A/D converter 4 channel D/A output 16 digital I/O lines 1 counter/timer 12-bit resolution 25 μs conversion time
EuroBEEB, CELESTE	Control Universal Ltd, CAMBRIDGE CB5 5QF	16 analogue input channels 1 analogue output channel 16 digital I/O lines with 4 control lines 8-bit resolution 100 μs conversion time
MICROLINK	Biodata Ltd, MANCHESTER M8 8QG	Modular data acquisition and control system for IBM-PCs and compatibles

Peripheral systems for real time data acquisition and control tend to be quite expensive with total costs sometimes in excess of £1000.00. With this level of investment, these systems are often tailored to service a specific application. They are often therefore dedicated to some research type investigation and are less commonly applied in a purely general sense. An application featuring the EuroBEEB system is described in section 10.3. The applications software which is normally supplied with a peripheral system is usually of a simple illustrative nature. Most users would require to write their own programs to suit their particular requirements.

A recent extension to peripheral systems is the so-called 'virtual instrument concept'. This is a technique where the graphics capabilities of the microcomputer are used to simulate the analogue instrumentation which is normally encountered in measurement systems.

6.8   THE VIRTUAL INSTRUMENT CONCEPT

In conventional measurement and control applications, visual displays are obtained in the familiar range of traditional instrumentation. Typical instrumentation, for example, includes analogue and digital meters, oscilloscopes and chart recorders. The concept of the virtual instrument is to use a microcomputer to measure the physical data and to display the data on a screen in a form recognisable to that which would be observed on the traditional instrumentation. The microcomputer, then, is programmed to mimic a range of analogue instrumentation.

Current commercial software normally facilitates a menu, or mouse selectable range of 'instruments' to be displayed at any one time. Various data processing can be selected with the results similarly displayed, or a complete experimental set-up may be simulated on the screen. The experimental mimic would include all of the associated 'instrumentation' suitably displayed and actively responding to change in the measured variables. The virtual instrument concept is a relatively new idea in microcomputer-based measurement systems and it can be expected that fairly rapid developments will ensue in the future. The obvious advantage offered by virtual instrumentation is the possibility of negating the need for real analogue instrumentation and the inevitable high cost associated with the traditional approach.

A practical example of a simple virtual instrument is given in section 9.6. This is followed, in section 9.7, with two further examples using the virtual instrument concept in applications to system monitoring.

REFERENCES

3D Digital Design and Development, *INLAB/THINKLAB, Modular Versatile Real Time Data Acquisition and Control System*, 3D Digital Design & Development Ltd, Southgate, London, 1987.

Barney, G. C., *Intelligent Instrumentation*, Prentice-Hall International, 1985.

Cassel, D. A., *Microcomputers and Modern Control Engineering*, Reston Publishing Co., Prentice-Hall, 1983.

Grant Instruments, *The 1200 series Squirrels*, Grant Instruments (Cambridge) Ltd, Barrington, Cambridge, 1986.

Kraus, T. W. and Myron, T. J., *Self-tuning PID controller pattern recognition approach*, Control Engineering, June 1984.

Milne, J. S., Fraser, C. J. and Gardiner, I. D., A low-cost data acquisition and control system based on a BBC microcomputer, *Conf. on Developments in Measurements and Instrumentation in Engineering*, Mechanical Engineering Publications, 1985.

Morrison, R., *Instrumentation Fundamentals and Applications*, Wiley, 1984.

National Instruments, *LabWindows Release 1.1 for the IBM-PC Series*, National Instruments, Kingston upon Thames, 1988.

Peter Nelson Design Consultancy, *OASIS Virtual Instrument System*, Peter Nelson Design Consultancy, Norwich, 1987.

Zeigler, J. G. and Nichols, N. B., Optimum setting for automatic controllers, *Trans. ASME*, 1942.

Chapter 7
Data Acquisition with the BBC Microcomputer

7.1 THE ADVAL COMMAND IN BBC BASIC

The BBC microcomputer comes already equipped with on-board, 12-bit A/D conversion capabilities. The converter can accept analogue inputs in the range 0 to 1.8 volts. With 1.8 volts input, the converter, which is the integrating type, will produce the number 65520. This is a 12-bit increment of 16 and the normal 12-bit range of 0 to 4095 can be obtained by DIViding by 16. Software access to the A/D converter is provided by the BASIC keyword ADVAL(n) where n refers to the channel number 1, 2, 3 or 4. Thus to read an input on channel 3, the required program is:

    100 I = ADVAL(3)
    110 PRINT I

The above will return a number between 0 and 65520 depending on the magnitude of the input, between 0 and 1.8 volts, on channel 3. To convert this to an equivalent 12-bit number, the coding could be altered to:

    100 I = ADVAL(3) DIV 16
    110 PRINT I

Alternatively, for an equivalent 8-bit number, line 100 requires to be altered to:

    100 I = ADVAL(3) DIV 256

The DIV command gives the whole number part of the result of a division.

Although there is only one A/D converter on-board the BBC machine, four analogue inputs are accommodated through multiplexing. Thus when an ADVAL command is implemented, all four channels are sampled consecutively in reverse

141

order. The sampling and conversion time for one channel is about 10 ms so that the cycle time in sampling the four channels is of the order of 40 ms.

To limit the sampling to specific channels, the command:

*FX16, m

can be used.

Thus *FX16, 1 will enable channel 1 only and the maximum sampling rate of about 100 Hz can be approached. Note however that if channel 3 is enabled (i.e. with *FX16, 3), then channels 1 and 2 will also be sampled in sequence and a maximum scan rate of about 33 Hz only is possible. The conversion rate can be further improved by switching the A/D converter from 12-bit mode to 8-bit mode. This is done with the command:

*FX190, 8

This will speed up the sampling rate but the resolution is reduced.

The analogue input terminal on the BBC microcomputer, a standard DIN plug connector, is depicted in figure 7.1.

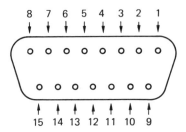

Figure 7.1   BBC microcomputer analogue input plug

The pin functions are:

1	+5 V	9	LPSTB
2	0 V	10	PB1
3	0 V	11	V_{ref}
4	CH3	12	CH2
5	Analog Gnd	13	PB0
6	0 V	14	V_{ref}
7	CH1	15	CH0
8	Analog Gnd		

Note that channel 1, accessed by ADVAL(1), is CH0 on the input pin.

For a single uni-polar input, the signal can be connected to pins 15 and 8, and an example program to read in 1000 values at channel 1 (CH0 on the pin) is given below:

```
 5 *FX16,1                        enable only channel 1
10 DIM I(1000)
20 FOR K = 1 TO 1000
30 I(K) = ADVAL(1)               data acquisition loop
40 NEXT K
50 FOR K = 1 TO 1000
60 PRINT (I(K) DIV16)*1.8/4095 calibration and print loop
70 NEXT K
```

To check the sampling rate, the program can be extended to include:

```
 5 *FX16,1
10 DIM I(1000)
15 TIME = 0
20 FOR K = 1 TO 1000
30 I(K) = ADVAL(1)
40 NEXT K
42 T = TIME
44 PRINT"Sampling time = ";T/100;" seconds"
```

At line 15, the internal timer of the microcomputer has been set to zero. The timer counts in one-hundredth of a second intervals and at line 42 the variable T will be assigned the value of the elapsed time in 1/100 seconds since line 15. In other words, the time taken to complete the data-acquisition loop. In running the program it is found that T = 3.75 seconds and that 1000 readings have been sampled in this time. This gives a single sample time of 3.75 ms which is faster than the sampling and conversion time of the system. If the input was say a half rectified sinusoid of 1.5 volts peak and 10 Hz frequency, then a section of the digitised output has the form shown in figure 7.2.

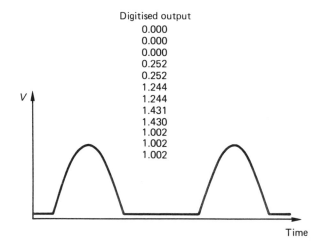

Digitised output
0.000
0.000
0.000
0.252
0.252
1.244
1.244
1.431
1.430
1.002
1.002
1.002

Figure 7.2   Digital representation of an analogue signal

It is clear that all is not well since the digitised output appears erratic and bears only a passing resemblance to the input signal. What in fact has happened is that the sampling rate is too fast and there is not enough time between samples to complete the A/D conversion properly. The program therefore requires some check that the conversion has been completed and this is provided in BBC BASIC with the statement:

REPEAT:UNTIL ADVAL(0) = 256

ADVAL(0) returns the number of the last channel to complete conversion. If the value returned is zero, then no channel has completed conversion. By qualifying this statement within a REPEAT UNTIL loop, the program is effectively halted until a value is returned, at which time the conversion is complete. Incorporating this modification into the data-acquisition loop gives:

```
20 FOR K = 1 TO 1000
25 REPEAT:UNTIL ADVAL(0) = 256
30 I(K) = ADVAL(1)
40 NEXT K
```

The output generated with this addition is:

```
0.000
0.471
1.296
1.458
0.901
0.000
0.000
0.000
0.000
0.610
1.351
1.434
0.820
0.000
0.000
```

This is a much closer representation of the input signal.

In addition, the total sampling time is 11.45 seconds, or 11.45 ms per sample, which is safely outside the conversion time of the A/D converter. The above, in fact, is close to the fastest rate at which an analogue signal can be sampled using the 12-bit on-board A/D converter of the BBC microcomputer. An improvement in speed can be obtained by switching to 8-bit mode in which the sampling rate might be increased to approximately 250–300 Hz.

To increase the sampling rate, assembly language data-acquisition routines must be used in conjunction with a fast A/D converter, interfaced to the user port. The basic requirement is given diagrammatically in figure 6.5.

It can be seen that in addition to the 8-bit data transfer lines to the computer, there must also be two other additional control lines to:

(i) send a start conversion signal to the A/D converter
(ii) sense the end of the conversion.

It becomes necessary, at this juncture, to consider some additional features of the BBC microcomputer user port.

7.2 THE BBC MICROCOMPUTER USER PORT

The pin functions on the user port are as given in figure 7.3.

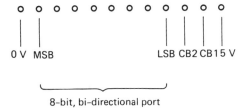

0 V MSB   LSB CB2 CB1 5 V

8-bit, bi-directional port

Figure 7.3 BBC microcomputer user port

Supplementary to the 8-bit port and the two reference voltages are two additional lines, CB1 and CB2. These latter two lines are control lines available in the 6522 VIA, see section 5.3.

Although the 6522 VIA is a complex multi-functioned chip, there are only two internal 8-bit registers of significance to data-acquisition applications. These are the Peripheral Control Register (PCR), and the Interrupt Flags Register (IFR).

The PCR, which is used to select the operational mode for the control lines, is arranged as follows:

Bit	7	6	5	4	3	2	1	0
Function		CB2		CB1		CA2		CA1

Bit 4 of the PCR selects the active transition of the input signal applied to CB1. Thus if PCR 4 was set to logic 1, the CB1 interrupt flag would be set by a positive transition, low to high, of the CB1 input signal. Setting PCR to logic 0 causes the same effect on the interrupt flag with a negative transition, high to low, of the CB1 input signal. The latter setting is therefore appropriate for sensing an end of conversion signal from an A/D converter; refer to figure 6.11.

The CB2 control functions are set by the state of bits 5, 6 and 7 of the PCR and eight variations are possible. With bits 7, 6 and 5 set to 1, 1 and 0 respectively, the CB2 line is set low, whereas with bits 7, 6 and 5 set to 1, 1 and 1, the CB2 line is set high. This in effect constitutes a 'start conversion' signal and can be used as such.

The IFR, which keeps a check on system and 6522 VIA related interrupts, is arranged as follows:

Bit	7	6	5	4	3	2	1	0
Function	IRQ	T1	T2	CB1	CB2	SR	CA1	CA2

The only flag of any consequence to the application in question is CB1. This flag will be set high by an active transition of the signal on the respective line and will be cleared by reading or writing to the user port address.

The location of the two registers in the BBC microcomputer are:

PCR – &FE6C
IFR – &FE6D

Hence to set up the control lines to operate an A/D converter, the following assembly language routine can be used:

```
 10 P%=&A00
 20 [
 30 LDA#0
 40 STA &FE62      \ user port set as input
 50
 60 LDA#&C0        \ start A/D converter pulse
 70 STA &FE6C      \ CB2 set low
 80 LDA#&E0
 90 STA &FE6C      \ CB2 set high
100
110 .loop
120 LDA &FE6D      \check for end of conversion on bit
130 AND#&10        \ 4 of the IFR,ie 00010000, while all
140 BEQ loop       \ other flags are masked
150
160 LDA &FE60      \ store converted value, also clears IFR
170 STA &4000      \ locate at &4000
180 RTS
190 ]
200 CALL &A00
210 PRINT ?&4000
220 GOTO 200
230 END
```

The equivalent routine in BBC BASIC is:

10 ?&FE62 = 0	Set the DDR for input
20 ?&FE6C = &C0	Set CB2 low
30 ?&FE6C = &E0	Set CB2 high
40 IF(?&FE6D AND &10) = 0 GOTO 40	Check for end of conversion
50 K = ?&FE60	PEEK port and store result

7.3   FAST CONTINUOUS DATA ACQUISITION ON THE BBC MICROCOMPUTER

In almost all instances data sampling is, of necessity, an ongoing process. For an application where 100 samples of a transient variable are to be captured as quickly as possible, the on-board A/D converters of the BBC machine may not be fast enough and alternatively an external A/D converter may be interfaced to the user port. Using BBC BASIC, the data-acquisition program could take the form shown below:

```
10 DIM I(1000)
20 ?&FE62=0
30 FOR K=1 TO 1000
40 ?&FE6C=&C0:REM start conversion
50 ?&FE6C=&E0:REM CB2 low then CB2 high
55 REM check for EOC with CB1 going high
60 IF (?&FE6D AND &10)=0 GOTO 60
70 I(K)=?&FE60
80 NEXT K
```

In timing the FOR–NEXT loop it is found that the 1000 samples are acquired in 5.66 seconds, i.e. the sampling rate is equivalent to 176 Hz.

To improve on this rate, the data-acquisition loop must be written in assembly language. The following program will read in 250 values from an A/D converter. The program incorporates the essential elements of:

(i) sending a start conversion signal to the A/D converter
(ii) checking for the end of conversion
(iii) committing the data to storage
(iv) incrementing the storage byte location for the next successive reading.

```
10 N=&4000-1
20 DDR=&FE62:PORT=&FE60
30 PCR=&FE6C:ifr=&FE6D
40 FOR pass=1 TO 3 STEP 2
50 P%=&A00
60 [OPT pass
70 LDX#0          \zero counter
80 LDA#0          \user port set as input
90 STA DDR
100 .repeat       \data acquisition loop
110 LDA#&C0       \start A/D converter
120 STA PCR       \CB2 set low
130 LDA#&E0
140 STA PCR       \CB2 set high
150 .loop         \check for EOC, on CB1
160 LDA ifr       \using bit-4 of the IFR
170 AND#&10
180 BEQ loop
190 LDA PORT      \read A/D value and clear IFR
200 STA &4000,X   \store value at &4000+X
210 INX           \increment counter
220 CPX#&FA       \check for 250 values read
230 BNE repeat
240 RTS           \return to BASIC
250 ]
260 CALL &A00
270 END
```

The following FOR–NEXT loop can access the data in BASIC at some later stage in the program:

```
490 N = &4000 − 1
500 FOR K = 1 TO 250
510 N = N + 1
520 S(K) = ?N
530 NEXT
```

The above loop will assign the 250 data samples to each consecutive element of a 250 element array called S. Further manipulation and processing on the data can then conveniently be done in BASIC.

To time the data-acquisition routine, reference must be made to the number of clock cycles that each operation takes and the internal clock operating frequency. The BBC microcomputer's internal clock is set at 2 MHz so that any operation requiring say 4 clock cycles will take 2 μs for completion. The data sampling loop begins at line 100 and ends at line 230.

The timing is as follows:

```
100 .repeat
110 LDA#&C0          2 cycles      1.0 µs
120 STA PCR          4 cycles      2.0 µs
130 LDA#&E0          2 cycles      1.0 µs
140 STA PCR          4 cycles      2.0 µs
150 .loop
160 LDA ifr          4 cycles      2.0 µs
170 AND#&10          2 cycles      1.0 µs
180 BEQ loop         2 cycles      1.0 µs
190 LDA PORT         4 cycles      2.0 µs
200 STA &4000, X     5 cycles      2.5 µs
210 INX              2 cycles      1.0 µs
220 CPX#&FA          2 cycles      1.0 µs
230 BNE repeat       3 cycles      1.5 µs
```

Lines 150 to 180 constitute the end of conversion check and the number of times that this loop is cycled will depend on the conversion time of the A/D converter. If the A/D converter has a conversion time of say 25 μs, then the loop will be cycled six, or perhaps seven, times. This section of the program may then take a total of 28 μs for completion. The rest of the program is executed only once for every data sample taken and this takes up 14 μs, giving a total time for each sample of 42 μs. The corresponding sampling frequency is of the order of 24 kHz. The largest expenditure in time occurs in the check for the end of conversion and this is the most obvious area in which to improve on the sampling frequency. If an A/D converter having a conversion time of 1 μs was to be used, then only one trip round the loop from lines 150 to 180 would be required and the total sample time would then be 18 μs, theoretically giving a sampling frequency of 55 kHz. Using a 5/1 criterion for the avoidance of aliasing error, this gives a maximum signal frequency of about 11 kHz which covers a useful range of both physical and mechanical systems. An application to an internal combustion engine indicator is described in section 10.2.

As an alternative to sensing the end of conversion, lines 150 to 180 may be replaced with a series of No Operation instructions (NOP). This instruction has no effect other than to use up 1 μs of time and the required number of NOPs can be substituted for the end of conversion check.

The example given was restricted to the acquisition of 250 samples only and at a sampling rate of 24 kHz, the total sampling interval would be 10.4 ms. If the physical phenomenon being measured is quasi-steady, or intermittent in nature, then the sampling may be required to be carried out over a much longer period. The 250 restriction on the sample number is a consequence of the method used to index the data location bytes, i.e. lines 200, 210. This is called 'direct indexing' and uses the X register to reference each successive byte location for the storage of the data. As the X register is an 8-bit register, then it can range between 0 and 255 and can therefore reference up to 256 byte locations, i.e. one page of memory. If more than 256 samples are required, then some means of referencing both the page and the byte locations becomes necessary. One method is to use both the X and the Y registers for this purpose. The use of these registers is illustrated by extending the program to read in 1000 values at the maximum sampling frequency.

Note: 1000 samples = (1000/256) = 3 pages and 232 bytes remaining.

As a precursor, the 'base-address', or starting byte location for the data is stored at the zero page addresses of &70 and &71. This can be done in BASIC before the assembly language routine is called, that is

```
150 ?&70 = &00
160 ?&71 = &40
```

The two program lines above set the two-byte location of &4000 as the base address. Any subsequent instruction, for example STA(&70), Y will store the accumulator value at the current two-byte address offset by Y. The low byte will be referenced by the contents at location &70 and the high byte by the contents at &71. The assembly language routine becomes:

```
10 DIM X(1000)
20 REM data to be stored from &4000
30 ?&70=&00
40 ?&71=&40
50 DIM G% 140
60 P%=G%
70 REM machine code routine to store 1000 values
80 [OPT 0
90 LDX #3          \store number of pages in X register
100 LDA #0         \set user port as input
110 STA &FE62
120 .NPAGES
130 LDY #0
```

```
140 .APAGE
150 LDA #&C0        \start A/D conversion
160 STA &FE6C
170 LDA #&E0
180 STA &FE6C
190 NOP             \delay to allow for end of conversion
200 NOP
210 NOP
220 NOP
230 NOP
240 LDA &FE60       \load ADC value
250 STA(&70),Y      \store at &4000+Y
260 INY             \increment Y
270 BNE APAGE       \repeat until page filled
280 INC &71         \increment the page number
290 DEX
300 BNE NPAGES      \complete 3 pages
310 NOP
320 LDY #0
330 .REST
340 LDA #&C0        \start A/D conversion
350 STA FE6C
360 LDA #&E0
370 STA FE6C
380 NOP
390 NOP
400 NOP
410 NOP
420 NOP
430 LDA &FE60         \store the rest of the data
440 STA(&70),Y
450 INY
460 CPY #&E8
470 BNE REST
480 RTS
490 ]
500 CALL G%
510 LOC=&4000
520 NUM=0
530 REM print out the 1000 values stored
540 FOR K=1 TO 1000
550 X(K)=?LOC
560 LOC=LOC+&01
570 NUM=NUM+1
580 PRINT X(K),NUM
590 NEXT
600 END
```

If more than three pages are required then line 90 of the program can be altered to suit. The absolute upper limit is 255 pages with 256 data samples stored on each page. This would require almost 64K of memory for storage and obviously could not be accommodated on earlier BBC machines such as the model B, which has only 32K of memory available.

EXERCISES

1. Investigate the scan rate for reading a single channel, say channel 1, on the BBC microcomputer when working in:
   (a) 12-bit mode

(b) 8-bit mode.   [approx. 100 Hz; approx. 250 to 300 Hz]
Note: store the data as 16-bit array elements and write the data-acquisition
loop for maximum sampling rates with no calibration calculations or print
out.

Investigate also the magnitude of the input signal level on the sampling rate.

2. Using the reference voltage of the onboard A/D converter in the BBC microcomputer, pin 14 on the analogue plug, connect up a potentiometer and feed
the output to channel 1, pin 15, as shown in the diagram below:

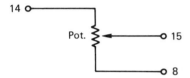

Write a BASIC program to sample channel 1 continuously and, while adjust-
ing the potentiometer, print the calibrated digital voltage to the monitor
screen.

3. Write a program in BASIC for 'fast data acquisition', using the analogue input
port. The data should be captured over a specified time interval and stored as
a one-dimensional array.

If available, use a waveform generator to produce a rectified, i.e. positive
only, sine function of amplitude 1.0 volts. Compare the captured digital
signal with the analogue input as the input frequency is increased from 10 Hz
to about 300 Hz.

REFERENCES

Bannister, B. R. and Whitehead, M. D., *Interfacing the BBC Microcomputer*,
    Macmillan Microcomputer Books, 1985.
Bray, A. C., Dickens, A. C. and Holmes, M. A., *Advanced User Guide for the
    BBC Microcomputer*, Cambridge Microcomputer Centre, 1983.
Bright, B., *Pocket Guide – Assembly Language for the 6502*, Pitman, 1983.
Opie, C., *Interfacing the BBC Microcomputer*, McGraw-Hill, 1984.
White, M. A., *Good BASIC programming on the BBC microcomputer*, Macmillan,
    1984.
Zaks, R., *Programming the 6502*, Sybex, 1980.

Chapter 8
IBM-PC and Compatibles

8.1 16-BIT MICROCOMPUTERS

In preceding chapters the fundamentals of microcomputer-based instrumentation have been addressed. Many, but not all, of the illustrative example programs are particular to the BBC microcomputer using the 6502 microprocessor. While many of the principles covered (e.g. number systems, transducers, signal conditioning and interfacing) are machine independent, cognisance must properly be given to the recent ascendancy of the 16-bit microcomputer.

The dominance of the International Business Machine, IBM, company in the 16-bit microcomputer market has led to the establishment of an industry standard machine in the form of the IBM-PC. In recent times, various clones of the IBM-PC have been launched and perhaps the most well known of these in the UK is the AMSTRAD PC 1512 or PC 1640. The PC 1640, introduced in 1988, is the most recent.

The AMSTRAD PC 1640 is fully compatible with the IBM-PC and can run all of the popular IBM business software. The PC 1640 is however a much less expensive machine and a typical system including colour monitor, 20 Mb hard disc drive and 360K double sided double density floppy disc drive can be obtained for under £1000. The additional computing power available with the 16-bit machine is quite extensive. The PC 1640 provides the user with 640K of RAM (compare this with the 128K of total memory available on the BBC Master), and the system is supplied as a complete stand alone unit. Predominantly a business machine however, the PC 1640 has limited input/output facilities and comes supplied solely with an RS232 serial port and a parallel printer port. To function as the workhorse in a data-acquisition/control context, the machine must be extended through the addition of appropriate expansion boards. Fortunately, there exists a large selection of these boards which can plug directly into any one of the PC's three option slots. A typical composite board might include 16 analogue input lines, 4 analogue output lines, 24 programmable I/O lines and 3 counter timers. The costs of these boards are variable, depending on speed, resolution and utility.

Early versions of the IBM-PC used the Intel 8088 microprocessor chip which had an 8-bit external data bus and essentially formed a compromise between an 8-bit and a 16-bit microprocessor. IBM adopted this strategy to take advantage of the extensive range of existing 8-bit ICs which were available on the market at that time. More recent machines, including compatibles like the AMSTRAD PC 1512 and PC 1640, use the Intel 8086 chip which is a full 16-bit processor.

The bus structure, see section 5.4, within the computer architecture enables the transfer of data on the 'data bus'. Of equal importance is the 'address bus' which updates the memory management circuitry relating which section of memory the data can be accessed from, or transmitted to. Obviously, with a larger bus, more information can be transmitted in any given time. The 8-bit/16-bit distinction normally refers to the data bus and generally defines the basic discrete chunk of data that the microprocessor can handle at any particular instant. The size of the data bus is of less importance, however, than the amount of memory that the microprocessor can access. This in turn is related to the size of the address bus. 8-bit microprocessors normally have a 16-bit address bus and this can allow the processor access to 2^{16} = 65536 memory locations (i.e. 64K). The 8086 also has a basic address bus of 16 bits, but can utilise an additional four lines so that it effectively has a 20-bit address bus. The resultant theoretical memory space equates to 1048K, or one megabyte. In practice, because of software limitations, most PCs with the 8086 microprocessor have an upper limit of 640K of RAM available to the user. This is still an extensive amount at memory, and is the biggest single advantage that 16-bit machines have over their 8-bit predecessors.

8.2 BASIC ON 16-BIT MICROCOMPUTERS

An added advantage associated with 8086-based microcomputers is that they all use the Microsoft-disc operating system, MS-DOS. In consequence, there is a vast amount of commerical software available which can run directly on 8086-based machines. This includes the well known word processors, spread sheets, databases and accounting software as well as the popular finite element analysis packages and CAD software. MS-DOS also means that an extensive range of the popular computer languages can be loaded into the machine and run. At the time of writing, these languages are prolific. They include for example Ada, Forth, Lisp, C and Pascal, among others. All have their adherents who readily extol their favourite language's virtues. It is our contention however that BASIC will remain popular since it is the language which most beginners are given a first exposure to. Modern versions of BASIC are well structured and Microsoft's

'QUICKBASIC', for example, is available as a compiled language which runs from four to ten times faster than any of the interpreted forms. This feature virtually negates all criticisms of the lack of speed inherent in interpreted BASIC.

Data-acquisition and control applications may be classified as either fast or slow. For slow acqustion rates, interpreted BASIC would normally suffice. Faster acquisition rates up to about 1 kHz might be achieved with compiled BASIC, but the very fast rates required for transient analysis would normally invoke the use of assembly language routines. One final point concerning BASIC is that, although many different dialects exist, the differences between each version are fairly minimal. Familiarity with one particular version of BASIC can readily be transposed to another.

The IBM-PC high-level BASIC language is 'BASICA', and in general BASICA cannot be run on IBM compatibles. Similar versions of this language including GWBASIC, XBASIC and PBASIC, which are clones of BASICA, will however run on the IBM machine as well as on a compatible. The essential elements of BASIC have been covered in chapter 4 and there is little advantage to be gained in detailing the minor variations which exist between the many versions of the language. If the reader has occasion to use an IBM, or compatible, in a data-acquisition system, then fuller details on the appropriate BASIC may be obtained from the software vendors.

8.3 MS-DOS

MS-DOS is the operating system developed by Microsoft Inc. and, because of the influence of IBM, has become a general standard for 16-bit microcomputers. The commands available with this operating system are extensive and some of those, in more common usage, are reviewed in this section.

The most popular, current version of the operating system is DOS 3.3. For IBM-PCs and compatibles, DOS is usually contained on disc and must be loaded into the machine when the machine is first switched on. The initial response of the machine will depend on the peripherals attached, e.g. two floppy discs or one floppy disc and a hard disc. For a machine with two floppy discs, referred to as drive A and drive B, MS-DOS must be loaded in drive A on start-up. If the machine has a hard disc, containing MS-DOS, then on start-up, the machine will look at drive A, the floppy disc drive first. If there is no disc in drive A, the machine will automatically load MS-DOS from the hard disc, which is referred to as drive C. Following the loading of MS-DOS in this latter mode, the machine will indicate that drive C is the default drive and the monitor will display:

C>

If MS-DOS has been loaded from a floppy disc in drive A, the monitor will display drive A as the default drive, that is

A>

Changing from one drive to another is implemented by typing the required drive symbol, followed by a colon. For example:

to change from drive C to drive A:

system prompt	type
C>	A: (enter)

Note: (enter) or (return) is a single key function on the keyboard.

Once access to the required drive has been obtained, a list of all the files and sub-directories that are contained on the relevant disc may be obtained by typing:

DIR

The above command will display all the files currently stored on the disc. In addition, the filetype is also displayed. For example:

MYPROG.BAS
FORMAT.EXE
BACKUP.EXE
DISPLAY.COM
TEST.EXE

The file MYPROG.BAS indicates that MYPROG is a source code and .BAS specifies that it is written in BASIC.

The extension .COM indicates a machine code filetype, as does the .EXE filetype extension. Executable versions of programs can be run directly from MS-DOS. For example, to run the executable program TEXT.EXE from MS-DOS, the program name, without its extension, is simply typed in, followed by enter:

system prompt	type
A>	TEST

The second file listed, FORMAT.EXE, is a machine code program, usually supplied with the hardware, and is used to format blank discs.

MS-DOS provides additional software facilities to:
1. Format or copy a disc.
2. Copy, delete or rename a file.
3. Determine the size of a file.
4. Determine the available storage space remaining on a disc.
5. Use an Editor to create, or alter programs and text.

Format and copy commands will depend on the disc drives available. In the following examples it is assumed that the machine has both a hard disc and a floppy disc, so that the default drive will be the hard disc.

To format a blank floppy disc, place the unformatted disc in drive A and type:

FORMAT A:

Note: this command will also clear the disc in drive A of any information which it previously contained.

To copy the file MYPROG.BAS from drive A to drive C and rename the file as TEST.BAS, type:

COPY A:MYPROG.BAS C:TEST.BAS

To delete the file MYPROG.BAS, on the floppy disc, type:

DEL A:MYPROG.BAS

Many more commands are available in DOS and reference should be made to the appropriate user's manual for more explicit details. Whether running PC-DOS on an IBM, or MS-DOS on a compatible, the full power of the operating systems can only be appreciated through experience in their usage. Readers can only therefore be urged to experiment and familiarise themselves with the operating system commands at the computer keyboard.

8.4 DATA ACQUISITION AND CONTROL ON AN IBM-PC OR COMPATIBLE

For measurement and control applications the IBM-PC, or equivalent, has to be extended with the provision of user ports. A large variety of expansion boards is available for this express purpose and many provide other additional features. A typical example, supplied through Flight Electronics Ltd, is the PC SUPER ADDA-8.

The PC SUPER ADDA-8 incorporates 8-bit resolution on a 64 multiplexed A/D channel. The analogue input is bi-polar only, in the range −5 V to +5 V, and two software-controlled D/A lines are also available for control purposes. The board, in addition, provides 24 digital I/O lines and 3 independent 16-bit ripple counters. The conversion time of the ADC is specified as less than 25 μs, with a resolution to ± the least significant bit. The board also incorporates a switch selectable port address which can be configured to either 170–17F hex, or to 1F0–1FF hex. Channel selection may be software set in BASIC. For example, to monitor the input continuously at, say, channel 47, the following coding is required:

```
10 PORT = &H170        set base address
20 OUT PORT, 47        select channel
30 OUT PORT + 1, 0     start conversion
40 X = INP(PORT + 1)   read input at selected channel
50 PRINT X
60 GOTO 30
```

To scan continuously through the 64 channels consecutively, the coding can be altered to:

```
10 PORT = &H170            set base address
20 FOR I = 0 TO 63
30 OUT PORT,I              select channel
40 OUT PORT + 1,0          start conversion
50 VALUE = INP(PORT + 1)   read input at selected channel
60 PRINT "ON";I; "=";VALUE
70 NEXT I
80 GOTO 20
```

In the first line of both sample programs, the base address is set to the corresponding address selected on the jumper switches. The form of the instruction may differ from one A/D expansion board to another, but the formal 'procedure' will remain invariant. In high level BASIC there is usually no need to check for the end of conversion. The data-acquisition procedure is therefore:

1. define the card base address with PORT =
2. select an input channel
3. send out a start conversion signal
4. check for end of conversion, not usually required in BASIC
5. read the A/D converter
6. store data in memory.

Various looping routines can then be devised, as illustrated in the examples, to monitor continuously, or to scan through, the channels.

If the digital I/O lines are to be used in a control application then the ports must initially be configured for input or output as required. This is done by writing the appropriate numerical value to the control register, see also section 5.3.

Many other similar A/D expansion cards are available and a selection of some popular makes are given in table 8.1.

For slow to moderate speed applications, the routines given in BASIC may be perfectly adequate. The capture of fast transient states will almost certainly necessitate the use of assembly language routines.

Microcomputers based on the 8086 and 8088 microprocessor use the same instruction set. An assembly language routine written for an IBM-PC, with the 8088 processor, would run equally well on an AMSTRAD-PC, with the 8086 processor. The converse is also true, and from an assembly language programmer's point of view, the two machines are identical. The 16-bit technology which features in these two microprocessors nonetheless makes the assembly language routines slightly more complicated than the equivalent instructions for the 8-bit 6502 microprocessor. The essential details of the 8086/8088 instruction set for fast data-acquisition applications are covered in section 8.5.

The fundamental principles of digital control have been outlined in chapter 6 and these principles remain independent of the microcomputer hardware em-

Table 8.1   A/D, D/A expansion cards for IBM-PCs and compatibles

Card name	Manufacturer	Specification
ACM-44	Blue Chip Technology, Deeside, CLWYD, CH5 3PP	24 digital I/O lines 4 D/A output lines 16 single-ended A/D input channels, or 8 differential A/D input channels 1 timer for interrupt generation 8-bit resolution 2 μs conversion time link selectable input range
PC-26A	Amplicon Liveline Ltd, BRIGHTON, BN2 4AW	16 A/D input channels 12-bit resolution 25 μs conversion time uni-polar or bi-polar input ranges
PC-30A	Amplicon Liveline Ltd, BRIGHTON, BN2 4AW	Similar to the PC-26A but additionally including: 24 digital I/O lines 2 12-bit /DA output lines 2 8-bit D/A output lines 1 counter/timer 35 μs conversion time

ployed. On the IBM-PC, or compatible, the control functions are implemented through an expansion board using the I/O lines available. The SUPER ADDA-8 board, for example, provides 24 I/O lines. This constitutes an 8255 PPI (see section 5.3), with three 8-bit ports, A, B, and C, which can all be software set for input or output. The control examples which follow are based, in fact, on the Amplicon PC-30A expansion card. Any similar card may equally well have been used.

1. Sequential Control

Figure 6.16 shows the relevant pneumatic circuit and the input and output ports can be hardwired, in the same manner, to ports A and B on the expansion board. The flow diagram illustrates the control strategy as before. A PC may be fitted with say the Amplicon PC-30A expansion card which has two 8255 PPIs. The first PPI is associated with the ADC functions while the second PPI provides the three digital I/O ports. The base address of the card is 700 hex and the digital ports A and B have the hex addresses of 708 and 709 respectively. The appropriate BASIC coding for the sequential control of the pneumatic system is as follows:

```
10 REM PORTA INPUT AND PORTB OUTPUT
20 OUT &H70B,&H90
30 PORTA=&H708
40 PORTB=&H709
50 REM SWITCH AIR ON
60 OUT PORTB,&HFE
70 FOR K=1 TO 200:NEXT K
80 REM CYLINDER TO THE LEFT
90 OUT PORTB,&HFC
100 REM WAIT UNTIL END OF STROKE
110 IF (INP(PORTA) AND 1)=1 GOTO 110
120 REM SEND CYLINDER TO THE RIGHT
130 OUT PORTB,&HFA
140 REM WAIT UNTIL END OF STROKE
150 IF (INP(PORTA) AND 2)=2 GOTO 150
160 GOTO 90
```

2. ON/OFF Control

Figure 6.17 shows details of a system to be controlled using the PC-30A expansion card.

The analogue channel to be sampled is selected by writing to the upper nibble of PORTC and the conversion is initiated by toggling bit 0 of the lower nibble from low to high while maintaining bit 1 high. The 12-bit number corresponding to the physical temperature is read as the 8 LSBs from PORTA and the 4 MSBs from the lower nibble of PORTB, with the hex addresses of 700 and 701 respectively. Note that these ports are not the same as that associated with the digital I/O function. The A/D input and the digital I/O functions each have their own set of three ports A, B and C.

The program listing to achieve the ON/OFF control strategy specified in section 6.6 is:

```
10 REM SET DIGITAL I/O PORTS
20 REM PORTB OUTPUT
30 OUT &H70B,&H90
40 REM SET PORTS FOR 12-BIT ADC CARD
50 REM PORTA AND PORTB INPUT, PORTC OUTPUT
60 OUT &H703,&H92
70 TMIN=39.5:TMAX=40.5
80 REM HEATER OFF, FAN OFF
90 OUT &H709,255
100 REM READ WATER TEMPERATURE
110 REM START ADC ON CHANNEL 1
120 OUT &H702,&H12
130 OUT &H702,&H13
140 REM NO NEED TO CHECK FOR THE END OF CONVERSION
150 A=INP(&H700)
160 B=(INP(&H701) AND 15)
170 I=A+(256*B)
180 T=I*100/4095
190 REM FOR T<TMIN HEATER IS ON AND FAN IS OFF
200 IF (T<TMIN) GOTO 270
210 REM FOR T<=TMAX AND T>=TMIN
220 REM THEN HEATER IS OFF AND FAN IS OFF
230 IF (T<=TMAX) GOTO 90
240 REM FOR T>TMAX THEN HEATER IS OFF AND FAN IS ON
250 OUT &H709,253
260 GOTO 120
270 OUT &H709,254
280 GOTO 120
```

3. Closed Loop Feedback Control

Figure 6.21 illustrates the process to be controlled using a digital algorithm to implement a PID control strategy. A PC having the same expansion card as in the previous example is to be used. The value indicative of the required control effort is output from Port B of the digital I/O. This in turn is input to an external 8-bit DAC. The program listing follows:

```
10 REM SET UP DIGITAL I/O CARD
20 REM PORTB OUTPUT TO DAC FOR CONTROL POWER TO VALVE
30 OUT &H70B,&H90
40 REM SET UP PORTS FOR ADC CARD
50 REM PORTA AND PORTB INPUT, PORTC OUTPUT
60 OUT &H703,&H92
70 CLS
80 INPUT "LEVEL REQUIRED (MM)= ";SP
90 INPUT "GAIN (K)= ";K
100 INPUT "INTEGRAL TIME (Ti)secs= ";TI
110 INPUT "DERIVATIVE TIME (Td)secs= ";TD
120 DT=1:SUM=0:EP=0
130 T1=TIMER
140 GOSUB 1000
150 IF (TIMER-T1)<1 GOTO 150
160 GOTO 130
1000 REM READ LEVEL ON ADC CHANNEL 1
1010 OUT &H702,&H12
1020 OUT &H702,&H13
1030 A=INP(&H700)
1040 B=(INP(&H701) AND 15)
1050 I=A+(256*B)
1060 PV=1000*I/4095
1070 E=100*(1-(PV/SP))
1080 SUM=SUM+E
1090 OUTPUTI=(DT/TI)*SUM
1100 OUTPUTD=(TD/DT)*(E-EP)
1110 OUTPUT=K*(E+OUTPUTI+OUTPUTD)
1120 IF OUTPUT<0 THEN OUTPUT=0
1130 IF OUTPUT>100 THEN OUTPUT=100
1140 REM OUTPUT TO PORTB ON DIGITAL I/O CARD
1150 OUT &H709,(OUTPUT*2.55)
1160 EP=E
1170 RETURN
```

8.5 PROGRAMMING THE 8086/8088 FOR FAST DATA ACQUISITION

Before we can consider machine code routines for fast data capture, it is necessary to have an appreciation of the basic internal features of the 8086/8088 microprocessor. To programmers familiar with the 6502 microprocessor, the most apparent difference in the 16-bit chip is the number and range of the internal registers available. There are fourteen 16-bit registers as shown in figure 8.1.

The first group AX, BX, CX, and DX are 16-bit general-purpose registers. Each of these may be referenced as two separate 8-bit registers AH, AL, BH, BL etc. The left-most 8-bits of AX, for example, form the register AH, while the right-most 8-bits constitute AL. The AX register serves as a primary accumulator.

Data Registers

AX	AH	AL	Accumulator
BX	BH	BL	Base
CX	CH	CL	Count
DX	DH	DL	Data

Pointer & Index Registers

BP	Base Pointer
SP	Stack Pointer
SI	Source Index
DI	Destination Index

Segment Registers

CS	Code Segment
DS	Data Segment
SS	Stack Segment
ES	Extra Segment

PC	Program Counter

PSW	Status Word

Figure 8.1   8086/8088 internal registers

BX is referred to as the Base register and is used in the calculation of the memory addresses. CX is the Count register and is decremented by string and loop operations. CX may typically be used to control the number of times a repetitive sequence of instructions is performed. DX is termed the Data register and may be used to store the I/O address during certain I/O operations.

In 8086 assembly language, the mnemonic MOV is essentially the equivalent to load and store, see section 4.6, in 6502 assembly language. Thus the instruction:

MOV AX, 3

will assign the decimal number 3 to the AX register. To avoid ambiguity between hex and decimal numbers, the qualifier H or D, respectively, is appended to the right-hand end of the number:

MOV AX, 3D

The assembler, in fact, assumes that all numbers are decimal unless otherwise stated, such that the two forms of the instruction given above will perform the exact same function.

Consider the instruction:

MOV DL, AH

There is further ambiguity here since the instruction could be taken to mean either load the contents of register AH into register DL, or load the hex number A into register DL. To avoid any confusion, the general rule which applies is that all hex numbers beginning with a letter should be prefixed with a zero.

Thus the unambiguous instructions would be:

MOV DL,0AH	load the hex number A into DL
MOV DL, AH	load the contents of AH into DL

Memory addressing is accomplished via the base register, BX, for example:

MOV [BX], AX

The above instruction will store the contents of the AX register to an address given by the contents held in the BX register.

Storing data in memory may be done one or two bytes at a time. If the AX register contained 4AF2H and the instructions MOV [20H],AX were executed, location 20H would be assigned the number F2H and location 21H would contain 4AH. The actual process is a little more complex than this, but for the sake of simplicity we will leave a fuller description until later. 16-bit data is stored with the lower byte first, followed by the higher byte. If the data is moved in the other direction, from a memory location to a register, then the order is reversed so that a self-consistency is maintained. In the majority of cases, programmers need not concern themselves with this aspect.

Using the CX register as a counter, the following program will store, in consecutive locations from 3000H to 3063H inclusive, the integer numbers ranging between 1 and 100.

TABLE:	; name of assembly language routine
MOV AL, 1	; assign an initial value of unity to AL
MOV BX, 3000H	; set the base address for the data storage
MOV CX, 63H	; set the counter to 99 decimal
DAT:	; program label
MOV [BX], AL	; store current value contained in AL, to the current
	; location given by BX.
INC AL	; increment the contents of AL by one
INC BX	; increment the contents of BX by one
LOOP DAT	; return to label DAT and repeat
RETF	; return to the CALLing program

The loop instruction returns the program to the DAT label, where the process repeats. A LOOP instruction, in addition, automatically decrements the CX register such that the process will be continued until CX becomes equal to zero. In the above case, the loop will repeat 100 times since the initial value in CX was 99 decimal.

The assembly language routine may be CALLed from a BASIC program but must first of all be assembled and linked to produce a machine code file with a .COM filetype extension. The procedures for embedding assembly language inserts into a BASIC program are quite complex. A much easier option is to use some of the many commercial software packages available to produce object modules for both the BASIC programs and the assembly language inserts. These object modules can then be LINKed to produce machine code versions of the complete program and subroutines. This procedure, which we would recommend strongly, is explained in more detail in section 8.6.

The pointer and index registers are used for array type processing in assembly language routines and will not be considered further. The segment registers are more important however, as they are directly involved with the memory address-ing function. The accessible memory is divided up into segments, each of which contains 64K locations. Thus, each individual memory address consists of two parts, which are a 16-bit segment address and a 16-bit offset referenced to the start of the segment. Memory addresses are formed by shifting the segment register four bits to the left and adding the effective memory address, to give the actual address. If the segment address was, say, 030AH, then the instruction MOV AL,[200H] would refer to the actual address given by:

(i)	Shift segment address four bits to the left	030A0 H
(ii)	Add to this, the offset	200 H
	Actual address	032A0 H

In the above manner, 'MOV AL,[200H]' would load the contents at location 32A0H into the AL register — a little more complicated than was previously suggested.

The value of 030AH for the segment address is contained in the data segment register, DS, and this comes under the control of the operating system. When a program is run, the operating system decides where in memory to store the program and any data that it generates. This is implemented through the DS register which can be manipulated externally but can only be loaded indirectly. For example:

    MOV BX,0FA0H
    MOV DS,BX

The above would set the DS register to 0FA0 hex.

Since the operating system manipulates DS, external control of memory addressing may be better actioned through the extra segment, ES, register which can be handled independently of the operating system. It must be remembered however that when data is stored to, or fetched from memory relative to ES, an override prompt must always be included in the instruction. For example:

MOV ES:[BX],AL   store the current value in the AL register at the location given by BX, relative to ES

The two remaining registers are the flags register and the program counter. The flags register indicates the processor status and need only be considered when arithmetic and related operations are being performed. The program counter carries out the same function in any microprocessor, which is to hold the address of the next executable statement in the program.

In the previous section, a BASIC program was given for a data-acquisition application, using the PC SUPER ADDA-8 expansion board. The equivalent coding in 8086 assembly language, to read and store 64K of data from the 8-bit A/D converter, might take the form:

```
data:                   ;assembly language routine label
MOV DX,171H             ;load the value PORT+1 (see BASIC listing) into DX
MOV AX,3000H            ;load 3000 hex into AX and thus into ES
MOV ES,AX              ;ie starting address for data=30000 hex
MOV BX,00H             ;set BX to zero. This is used to address the data
                        ;relative to ES
MOV CX,0FFFFH          ;load FFFF hex into CX, ie maximum value possible in a
                        ;16-bit register
CONV_1:                 ;start of data acquisition loop
MOV AL,00H             ;assign a value of zero to AL
OUT DX,AL              ;start conversion, equivalent to OUT PORT+1,0 in BASIC
PUSH CX                ;push current value of CX to the stack
MOV CX,38H             ;load a new value to CX to time a delay loop
DELAY:                  ;delay loop to allow conversion to be completed
NOP                     ;LOOP instruction causes CX to be decremented
LOOP DELAY              ;and continues until CX=0
POP CX                 ;retrieve CX value from the stack for the main loop
IN AL,DX               ;read 8-bit value into AL, equivalent to INP(PORT+1) in BASIC
MOV ES:[BX],AL         ;store 8-bit value at relative address held in BX, note
                        ;extra segment override
INC BX                 ;increment BX to address next data
LOOP CONV_1            ;repeat until CX=0
RETF                    ;return to the CALLing program
```

In the program given, the CX register is used primarily to count down the 64K of data. CX is also used however, to time a wait for end of conversion delay. While the delay loop is executed then, the current value of CX must be temporarily stored on the stack and then retrieved following the delay loop. This is done using the PUSH and POP instructions as shown.

The delay routine consists of two instructions and can be timed by referring to the number of clock cycles required for each instruction:

	cycles
DELAY:	0
NOP	3
LOOP DELAY	17 with branch, 5 without branch

Thus one trip round the delay routine, with a branch at the end, involves twenty clock cycles. The AMSTRAD PC 1640 uses the 8086 microprocessor which operates at 8 MHz. If this particular machine is used, then the single trip is completed in 2.5 μs. The last trip, through the loop, with no branch at the end, takes 8 clock cycles. The LOOP instruction also decrements the CX register by one on each execution. The instruction immediately before the delay cycle assigns a temporary value of 38 hex, i.e. 56 decimal, to CX, so that the delay routine will be executed 57 times. In other words, the actual delay, in real time, is 141 μs. The delay can be adjusted by re-setting the initial value assigned to CX. The timing for the rest of the data acquisition routine is:

	cycles
CONV_1:	0
MOV AL,00H	4
OUT DX, AL	8
PUSH CX	11
MOV CX, 38H	4
Delay routine	1128
POP CX	8
IN AL, DX	8
MOV ES: [BX], AL	19
INC BX	2
LOOP CONV_1	17
Total	1209

At 8 MHz operating speed, 1209 clock cycles represents about 151 μs. The corresponding data-acquisition rate is therefore approximately 6.6 kHz. As with any microprocessor-based system, the biggest restriction on the sampling rate is associated with the delay routine required to allow for the completion of the A/D conversion. If a very fast A/D converter can be used, with say a 1 μs conversion time, the delay cycle need only be executed once. The total processing time would then take about 11 μs and a sampling rate of about 90 kHz could be realised. Note that the IBM-PC clock operates at 4.77 MHz and the sampling rates achievable on this machine are proportionally less than those on the AMSTRAD PC 1640.

The section of the program preceding the data-acquisition loop is concerned with the initialisation of the various registers, that is:

MOV AX, 3000H
MOV ES, AX

The above effectively sets the starting address for the storage of data at the location 30000 hex. This can be assured by the manner in which the machine calculates actual addresses, as previously explained.

The three remaining instructions in the initialisation process are:

```
MOV DX, 171H
MOV BX, 00H
MOV CX, 0FFFFH
```

The first assigns the numerical value of (PORT + 1) to the DX register. DX is subsequently used in the 'start conversion' and in the 'read port' instructions within the data-acquisition loop. The BX register is used to address the data storage byte locations relative to ES. BX is therefore initialised to zero and subsequently incremented by one with each cycle of the data-acquisition process. The CX register is set to FFFF hex, or 65535 decimal, and on every LOOP instruction, the CX register will be decremented by one. This ensures that the data-acquisition cycle will be performed 64K times and 64K individual data points will be committed to storage.

If more than 64K of data are required, then the program would have to be extended This would involve re-setting the registers as follows:

```
ES to 4000 H
CX to 0FFFFH
BX to 00H
```

The exact same data-acquisition loop could then be executed again and this would fill 64K of additional data between locations 40000H and 4FFFFH inclusive. A less cumbersome method is outlined in section 8.7.

The total RAM available is 640K of which a small amount is required by the operating system (0–500 hex). Uncertainty of the exact location of the code segment may prevent maximum use of memory for data storage. However, allowing three complete segments for the operating system and the program, i.e. setting the data starting address at 30000 hex, will still leave up to 7 segments, or 448K of memory for data storage. This would accommodate, for example, a 22 second sampling interval, at 20 kHz sampling rate, using 8-bit accuracy.

The program listing given is not the only means of performing the same operations. For example, repetitive loops may be devised using the extensive variety of compare, CMP, and jump, JMP, instructions available. It is beyond the scope of the present text to outline the range of alternatives possible, and the reader is directed to the reference texts at the end of the chapter and invited to experiment with variations to the coding given.

Access to the stored data following its capture may be done in BASIC, if speed is not essential. The assembly language routine is CALLed from a compiled BASIC program and a suitable program to be run in conjunction with the routine given might be:

```
10 REM Data retrieval
20 CALL data
30 DEF SEG = &H3000
```

```
40 FOR I = 0 to 65535
50 Y = PEEK (I)
60 PRINT Y
70 NEXT I
80 END
```

The program listing provides access to the data. The DEF SEG statment defines the address assigned to the appropriate segment register. Any subsequent command which accesses a memory location will access the location relative to the segment register. Thus by setting DEF SEG = &H3000, the data will be fetched from the same memory locations where it was stored in the assembly language routine. If the mean value of the data is required, then this may be calculated by adding the following lines of coding:

```
80 SUM = 0
90 FOR I = 0 TO 65535
100 SUM = SUM + PEEK(I)
110 NEXT I
120 AV = SUM/65536
130 PRINT AV
140 END
```

The data processing may be much more complex than the simple evaluation of the mean value. It must be pointed out however that with 64K, or more, of data to be handled, a BASIC version of the processing algorithm may be costly in terms of processing time. The obvious answer is to do as much of the processing as possible in assembly language subroutines. For example, to evaluate the mean value of the 64K of data captured in the data-acquisition routine, the following coding might be used as a much faster alternative:

```
SUM:                    ;routine label
MOV AX,00H              ;set AX to zero
MOV [0A000H],AX         ;set location A000 hex to zero
MOV [0A002H],AX         ;set location A002 hex to zero
MOV AX,3000H            ;load 3000 hex into AX and then into ES
MOV ES,AX               ;ie starting address for data =30000 hex
MOV BX,00H              ;set BX to zero
MOV CX,0FFFFH           ;set CX to FFFF hex
TOT1:                   ;summation label
MOV DL,ES:[BX]          ;assign 8-bit value at relative address held in BX
                        ;to the DL register
MOV DH,00H              ;set DH to zero
MOV AX,[0A000H]         ;load AX with current value in A000 hex
ADD AX,DX               ;add current value in DX to AX
MOV [0A000H],AX         ;assign current value of AX to location 0A000 hex
MOV AX,[0A002H]         ;load AX with current value in A002 hex
ADC AX,00H              ;add with carry 8-bit zero to AX
MOV [0A002H],AX         ;assign current value of AX to location A002 hex
INC BX                  ;increment BX
LOOP TOT1               ;repeat until CX=0
RETF                    ;return to the CALLing program
```

In the summation routine above, the registers are initialised to the same starting values they had in the data-acquisition program. Counting is controlled through the CX register and the data is indexed through BX relative to ES as before. Each piece of the data however is an 8-bit number which could conceivably have the maximum numerical value of 255 decimal. If every single piece of data had the maximum numerical value, then the summation procedure would return a total value of 255×65536 (or $2^8 \times 2^{16}$) = 16.71×10^6. A number of this size can only be accommodated as a 24-bit binary equivalent. This is allowed for in the program by storing the cumulative summation as a 32-bit number. The lower 16 bits of the number are addressed at A000H and the higher 16 bits addressed at A002H. The summation is performed within the loop labelled by TOT1. The first two instructions load the DX register with one element of the 8-bit data. Note that the high byte, DH is set to zero. The current lower 16-bit total is then loaded into AX and the contents of DX are added. At some point in the summation procedure, this total will exceed the maximum possible with a 16-bit number and the carry flag will be set to '1'. The current value in AX is then stored at A000 hex, whether the carry flag is set or not. Following this, the higher 16 bits of the total, located at A002 hex, are assigned to AX and zero is added to this total with a carry. The value stored at A002 hex will only therefore be incremented when a carry over from the previous ADDition has occurred. If this does happen then the ADC instruction will clear the carry flag and the summation will continue in a similar manner until the lower 16 bits of the cumulative total again set the carry flag.

Using the above routine, the 64K of data are summated and stored as a 32-bit number. The mean value can then be calculated in BASIC using the following coding:

```
80 CALL SUM
90 SUM1 = PEEK(&HA000)
100 SUM2 = PEEK(&HA001)
110 SUM3 = PEEK(&HA002)
120 SUM4 = PEEK(&HA003)
130 TOTAL = SUM4*(2^24)+SUM3*(2^16)+SUM2*256+SUM1
140 AV = TOTAL/65536
150 PRINT AV
160 END
```

The variables SUM1 to SUM4 are the four, 8-bit components of the 32-bit total.

When dealing with massive arrays of data, assembly language may prove to be essential in order to keep the data manipulation times down to reasonable levels. In the example given, the evaluation of a mean value of 64K of data takes just over 3 minutes in interpreted BASIC, 48 seconds in compiled BASIC, but less than two seconds with the assembly language insert.

8.6 COMPILED BASIC, ASSEMBLERS AND LINKERS

Programs written in BASIC can be run considerably faster if they can be converted via a compiler, or compiler and linker, into a machine code executable
version. A number of compiler programs are available, for example, Microsoft's
'QUICKBASIC' or Borland's 'TURBOBASIC'.

Compilers are special machine code programs which translate a BASIC program or other source code, into an object code. A linker is another machine
code program which can link specific object code with a variety of library
routines to produce a directly executable machine code. The compiler and linker
programs can be used therefore to produce much faster machine code versions of
programs written in high level BASIC. The machine code version of the program
is identified with a .EXE filetype extension and this program can be run directly
from MS-DOS.

If a BASIC source code called TEST.BAS is available on the directory, along
with the QUICKBASIC compiler program, an object module may be created
with the command:

BASIC TEST

There is no requirement to include the source code filetype extension, but
the command BASIC and the filename must be separated with a space. Following
this instruction, a check on the directory will show that a new file, TEST.OBJ,
has been created and saved. This carries the proviso that there were no errors
detected during the compilation. If the source code had contained errors, then
the compiler would report these and display them on the monitor screen. All
errors must obviously be corrected before a successful compilation can be
completed.

The translation of the object code into executable machine code is done with
the following command:

LINK TEST

Again note that there is a space between the command and the filename. A
further check on the directory at this point will show that another file, TEST.
EXE, has been created and saved. This latter file is the machine code version of
the original BASIC program and it may be run from MS-DOS by simply typing
the filename, without the filetype extension, that is:

TEST

On average, the executable version of the program will run about four to ten
times faster than the interpreted high level BASIC version.

Assembly language programs, or inserts which are CALLed from a BASIC
program, can similarly be assembled into an object module and this may be
done through the 'A86', or other macro-assembler program. Having written the

assembly language routine on a text editor, the object module is created with the command:

A86 +o filename.filetype

In the above command, the spaces on either side of the characters "+o" are essential. Note also that the character following the + is the lower case letter o, not zero. The rest of the command specifies the filename, complete with its filetype extension. The A86 assembler program will report any error messages and if none are detected, an object module will be created and saved to the directory.

If a BASIC program CALLs an assembly language routine during its execution, then the two separate programs must first of all be LINKed using the linker program. This is achieved with the command:

LINK program name + assembly language insert name

There is no restriction on the number of programs which may be linked, they are simply appended to the list and interspaced with a + character. The result of the LINK command is to produce a .EXE filetype extension of the complete program. The combined program may be run from MS-DOS by typing in the calling program name only.

The procedures outlined above can be summarised in a simple example.

BASIC source code:
```
TEST.BAS
        CALL TABLE
        FOR I = 0 TO 99
        Y = PEEK(&H3000 + I)
        PRINT Y
        NEXT I
        END
```

Assembly language insert:

```
TABLE.ASM
        TABLE:
        MOV AL, 1
        MOV BX, 3000H
        MOV CX, 63H
        DAT:
        MOV [BX] , AL
        INC BX
        INC AL
        LOOP DAT
        RETF
```

The first line of the BASIC program CALLs the assembly language insert labelled TABLE. The insert assigns the decimal numbers 1 to 100 inclusive to the memory locations 3000 hex to 3063 hex, relative to the DS register. The operating system will set the value assigned to DS and this need not be of further concern. On returning to the BASIC program, the FOR–NEXT loop accesses the data and prints it to the monitor screen.

The program given is of no particular relevance, but serves to summarise the steps involved in producing an executable version of the combined coding. The following instructions will perform this task.

BASIC TEST	; will create and save an object module ; of the BASIC program
A86 +o TABLE.ASM	; will create and save an object module ; of the assembly language insert
LINK TEST + TABLE	; will create an executable version of ; the complete program
TEST	; will run the complete program

The commands and procedures outlined are particular to the Microsoft QUICKBASIC compiler and object linker. If perhaps TURBOBASIC is to be used to create executable versions of source codes, then the set of commands are menu selectable. The series of commands used will also be dependent on the machine peripherals. Full details of the commands and procedures are normally supplied when buying the relevant software.

8.7 AN APPLICATION IN FLUID DYNAMICS

The 16-bit microcomputer lends itself well to the study of turbulent flows. Turbulence in boundary layer flows characteristically involves velocity fluctuations in the frequency range of 1 Hz to 2 kHz. The amplitude of the fluctuations are random and variable. In transitional states from laminar to fully turbulent conditions, the flow can, in addition, exhibit a number of quasi-steady features. A microcomputer-based data-acquisition system must therefore be capable of sampling rates of at least 10 kHz, to avoid aliasing errors. The system must also have the capability of storing a reasonable 'time' sample for meaningful analysis, say a minimum of ten seconds. These two prerequisites necessitate a minimum storage allocation of at least 100K, and preferably more. This requirement immediately excludes the BBC model B microcomputer and could be taxing on a BBC Master, even with 128K of memory available. The IBM-PC and a host of other 'lookalikes' can offer up to 640K of accessable memory and these machines are eminently suitable in accommodating the mass data storage required in the study of intermittently turbulent flows.

Using the ACM-44 expansion card described in table 8.1, a microcomputer-based system can be developed to satisfy the measurements requirements.

A suitable primary sensor is the standard DANTEC (formerly DISA) hot wire anemometer operated in the constant temperature mode. The anemometer output is non-linear and the most convenient means of signal conditioning may be effected using the standard DANTEC linearising unit. This unit allows switchable adjustment on zero offset, gain, exponent factor and many other parameters. An experienced operator can quickly implement a linear relationship between output voltage and fluid mean velocity. A typical calibration might be 0-2 V d.c., equivalent to 0-20 m/s mean velocity.

The other main variable in this type of experimental study is the position of the hot wire sensor relative to the solid surface over which the fluid flows. A possible solution here is to use a stepper motor-driven traversing mechanism. The stepper motor could be controlled and powered via some form of drive unit and this could be monitored to give a measure of the sensor position. The necessary hardware is readily available in standard DANTEC units and a block diagram of the instrumentation set-up is depicted in figure 8.2.

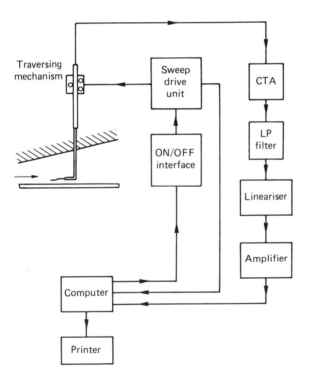

Figure 8.2 *Instrumentation set-up for studies of turbulent flow*

Also included in figure 8.2 is an ON/OFF interface which controls the power to the stepper motor drive unit and thereby effects control of the hot wire sensor position. The ON/OFF interface is shown in figure 8.3.

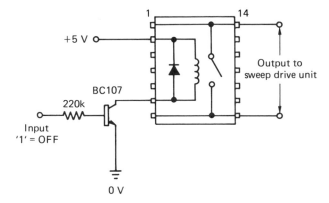

Figure 8.3 Stepper motor drive unit ON/OFF interface

The ACM-44 card provides 24 digital I/O lines and one of these, say the LSB of Port A, can be used to output a logic '0', or '1', corresponding to ON and OFF respectively. The voltage level indicated on the drive unit potentiometer can be monitored and translated, in software, to the actual hot wire sensor position.

The data-acquisition main program is given below:

```
REM Data Acquisition Program
REM
REM Initialisation Procedures

REM Set port address
PORT=&H300

REM D/A channel 0, used to output a supply
REM voltage for the ON/OFF interface
OUT PORT+4,100

REM Select channel 1, voltage output
REM from sweep drive unit
OUT PORT,1

REM Set the control word
OUT PORT+11,&H80

REM Output data to port A, a 1 will ensure
REM that the drive unit is switched off
OUT PORT+8,1

DIM B(40),Y(40),AV(40),HT(40)

REM Manual input data

PRINT
INPUT"Temperature in deg C - ";T
INPUT"atmospheric pressure in mm Hg - ";H
INPUT"y - datum in mm ";YDAT
INPUT"distance from leading edge in mm ";X
INPUT"enter the file name ";FILE$
```

```
PRINT
N=1
PRINT"    Y mm      Velocity m/s":PRINT

REM Read sensor position, average of 100 values

5
B(N)=0

FOR I=1 TO 100
REM Start conversion
OUT PORT+2,0
REM Read port, ie. the drive unit voltage
Y(N)=INP(PORT+2)
B(N)=B(N)+Y(N)
NEXT I
B(N)=B(N)/100

REM Data acquisition routine
REM Select channel 0, ie. the
REM the hot wire sensor voltage
OUT PORT,0

CALL ACQ

REM Mean value of signal

CALL TOTAL
C1=PEEK(&HA000)
C2=PEEK(&HA001)
C3=PEEK(&HA002)
C4=PEEK(&HA003)
AV(N)=(C4*(2^24)+C3*(2^16)+C2*256+C1)/(65536*2*12.5)

REM Re-select channel 1
OUT PORT,1
REM Start conversion
OUT PORT+2,0
REM Switch on drive unit
OUT PORT+8,0

N=N+1

10
REM Read port
Y(N)=INP(PORT+2)
REM Start conversion
OUT PORT+2,0
REM Check the step increment
IF (Y(N-1)-Y(N))<=4 GOTO 10
REM Switch off drive unit when increment is correct
OUT PORT+8,1
HT(N-1)=YDAT+(B(1)-B(N-1))*52.1/255
PRINT USING"   #.##        ##.###   ";HT(N-1);AV(N-1)
PRINT
IF N>10 GOTO 20

GOTO 5
20

REM Save data to floppy disc
```

```
OPEN"O",#1,"A:FILE$"
WRITE#1,T,H,YDAT,X,N
FOR I=1 TO N-1
WRITE#1,HT(I),AV(I)
NEXT I
CLOSE#1

STOP
END
```

The main program allows 20 sets of readings of sensor position and mean voltage to be recorded, with approximately equal increments in sensor position. The data capture and totalising subroutines are given below.

```
ACQ:                           TOTAL:
MOV CX,02H                     MOV CX,02H
MOV DX,302H                    MOV AX,00H
MOV AX,3000H                   MOV [0A000H],AX
START:                         MOV [0A002H],AX
PUSH CX                        MOV AX,3000H
MOV ES,AX                      START1:
MOV BX,00H                     PUSH CX
MOV CX,0FFFFH                  MOV ES,AX
CONV_1:                        MOV BX,00H
MOV AL,00H                     MOV CX,0FFFFH
OUT DX,AL                      TOT1:
PUSH CX                        MOV DL,ES:[BX]
MOV CX,24H                     MOV DH,00H
DELAY:                         MOV AX,[0A000H]
NOP                            ADD AX,DX
LOOP DELAY                     MOV [0A000H],AX
POP CX                         MOV AX,[0A002H]
IN AL,DX                       ADC AX,00H
MOV ES:[BX],AL                 MOV [0A002H],AX
INC BX                         INC BX
LOOP CONV_1                    LOOP TOT1
POP CX                         POP CX
MOV AX,4000H                   MOV AX,4000H
LOOP START                     LOOP START1
RETF                           RETF
```

In the subroutine ACQ:, 128K of data are captured and stored in two consecutive memory segments at memory addresses &H30000 to &H4FFFF inclusive.

Note how CX is used: (1) to count down the number of segments, (2) to count down 64K of data to each segment, and (3) to time the delay loop to allow for the end of conversion. The DX register is loaded with &H302, which is the value (PORT + 2). The ACM − 44 card is slightly different therefore from the PC SUPER ADDA-8 which uses (PORT + 1) to start conversion and read a channel. In all other respects, the data-acquisition and summing routines are essentially the same as those given in section 8.5.

Figure 8.4 shows a small section of the digitised signal captured using the above program. This particular sample shows a typical intermittently laminar and turbulent flow signal. Some post-processing is done on the data to identify turbulent patches and these are indicated with a shaded block drawn in below the signal.

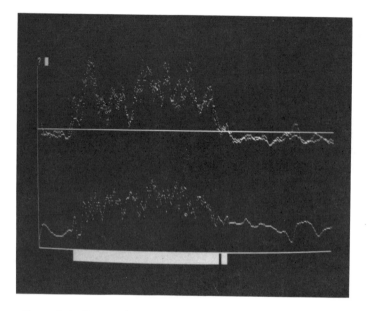

Figure 8.4 *Digitised signal from an intermittently laminar and turbulent flow*

The digitised signal may be further processed according to the requirements of the experimental investigation. Fuller details may be obtained from the references given at the end of the chapter. This is a very specialised research area however and need not be considered further. The data-acquisition methods which have been outlined nonetheless are general, and they may be equally well applied in the study of similarly complex transient physical phenomena.

EXERCISES

1. Using an assembly language insert, store the consecutive decimal numbers from 1 to 2000 in consecutive memory locations starting at &H3000. Read the numbers back and print them to the screen using a BASIC program.
   Note: numbers greater than 255 require to be stored as a 16-bit binary equivalent. For simplicity, it is easier to store all the numbers as 16-bit, in a high byte/low-byte form.
2. Write a program to calculate the average value of the data stored in exercise 1:
   (i) in BASIC
   (ii) in an assembly language insert.
   Compare the run times for each program.
   [6 seconds approx., less than 0.1 second]

3. Using any proprietary A/D expansion card for an IBM-PC, or compatible machine, write a program in BASIC to sample data from a single channel. The data should be stored as elements in a one-dimensional array.

Limiting the number of samples to 1000, use the TIMER function to determine the data acquisition rate.

If the software is available, compare the data-acquisition rate for a compiled version of the program.

4. Using an appropriate A/D expansion card for an IBM-PC, or compatible, write an assembly language insert to capture 64K of data at the maximum rate that the conversion speed of the card will allow.

If a waveform generator is available with control on input amplitude and frequency, test the data-acquisition routine for input frequencies up to the sampling frequency.

Note: if the input signal is periodic, then as the input signal frequency approaches the sampling frequency, the digitised data will tend towards a constant value.

REFERENCES

Amstrad PLC, *Amstrad PC 1640 – User Instructions*, Amstrad PLC and Locomotive Software, 1987.

Bate, J. S. J. and Burgess, R., *The Amstrad PC 1512. A User's Guide*, Collins, 1987.

Blue Chip Technology Ltd, *The ACM-44 Expansion Card for the IBM-PC*, Blue Chip Technology, Clwyd, 1988.

Borland International Ltd, *TurboBasic, IBM version*, Borland International, 1987.

Dantec Ltd, *Instruction manuals* (various), Dantec, Bristol, 1987.

Flight Electronics Ltd, *PC SUPER ADDA 8*, Flight Electronics, Southampton, 1987.

Fraser, C. J., Graham, D. and Milne, J. S., Fast data acquisition and digital signal processing in studies of turbulent flow, *Fourth International Conf. on Mech. Eng. and Tech., Zagazig Univ., Cairo*, 1989.

Graham, D., Fraser, C. J. and Milne, J. S., Digital measurements in intermittently turbulent flow, *AMSE International Conf. on Signals and Systems, Brighton*, 1989.

IBM Corp., *DOS 3.30 Reference (Abridged)*, 1st edn, IBM Corp. and Microsoft, 1987.

Liu, Y. and Gibson, G. A. *Microcomputer Systems: The 8086/8088 Family*, Prentice-Hall, 1984.

Microsoft Inc., *Quickbasic, Version 4.0*, Microsoft, 1988.

Rector, R. and Alexy, G., *The 8086 Book*, Osbourne/McGraw-Hill, 1980.

Ross, C. T. F., *BASIC Programming on the Amstrad 1512 and 1640 and IBM Compatibles*, Jonathan Ross, 1987.

Thorne, M., *Programming the 8086/8088 for the IBM and Compatibles*, Addison-Wesley, 1986.

Van Wolverton, *Quick Reference Guide to MS-DOS Commands*, Microsoft, 1987.

Chapter 9
Data Handling and Display

9.1 DATA INPUT AND OUTPUT

In high level BASIC, data may be input to a program using the interactive INPUT statement. For example:

    10 INPUT "NUMBER OF NODE POINTS = ";N

The semi-colon before the variable 'N' causes the cursor to stay on the same line that the text between the quotation marks will be printed on. This helps to keep the input tidy and 'user friendly'. Execution of the above statement will produce the following response on the monitor:

    NUMBER OF NODE POINTS = ?

The cursor will then remain next to the question mark until such time that a numerical value for N has been entered and followed with a return/enter.

The same effect can be produced with the statements:

    10 PRINT "NUMBER OF NODE POINTS = ";
    20 INPUT N

An alternative procedure for data input is to use the non-interactive READ statement in conjunction with an appropriate DATA statement. For example:

    10 READ N
       - - - -
       - - - -
       - - - -
    120 DATA 24

Notice that the DATA statement can appear anywhere, but it is considered good practice to place all DATA statements just before the END of the program.

If the data has to be re-read at some point, then the RESTORE statement can be used to re-set the data pointer. On execution of a RESTORE statement, the next READ statement in the program will access the first item of data in the first DATA statement which appears in the program. For example:

```
10 READ A,B,C
20 RESTORE
30 READ D,E,F
40 DATA 25, 30, 60
50 PRINT A,B,C,D,E,F
RUN
          25          30          60          25
          30          60
```

Any number of DATA statements can be used in a program. DATA statements may also be considered as one continuous list of items, irrespective of how many items are on the list. It should be remembered, however, that READ statements will always access the data in strict consecutive order.

Alpha-numeric data may be similarly handled. For example:

```
10 INPUT"ENTER TODAY'S DATE IN FORM (TUE/07/OCT/89) – ";A$
20 INPUT"ENTER TIME IN FORM (09:35 AM) – ";B$
30 INPUT"          TEST NUMBER          – ";TN
```

or non-interactively:

```
10 READ A$,B$,TN
    - - - - -
    - - - - -
    - - - - -
90 DATA WED/02/FEB/89,9:40 AM,17
```

Output of data, to a monitor or a printer, is accomplished with the PRINT statement. For example:

```
10 X = 4.5
20 A = X + 2
30 B = X/3
40 C = X*X
50 PRINT X,A,B,C
```

RUN
```
          4.5          6.5          1.5          20.25
```

When the PRINT items are interspaced with commas, each numerical item is printed to the right-hand side of a field of characters. The default width of the field is ten characters. Alpha-numeric data is printed to the left-hand side of the ten character field.

If we were to replace line 50 in the program with

    50 PRINT X;A;B;C

RUN
    4.56.51.520.25

the semi-colon delimiter causes the variables to be printed next to each other without spaces. The output is obviously confusing but the semi-colon allows for more control over the spacing of the printed output. For example:

    50 PRINT" ";X;" ";A;" ";B;" ";C

RUN
    4.5   6.5   1.5   20.25

The revised PRINT coding limits the spacing between each variable to two. This is imposed by inserting two spaces between each pair of the quotation marks. On the IBM-PC and compatibles, there is no need to insert spaces in the manner as shown above, because the BASIC interpreter accommodates this automatically. The technique however easily allows for the output of useful text along with the numerical data. For example:

    50 PRINT "X = ";X;", (X + 2) = ";A;", X/3 = ";B
    60 PRINT "and X squared = ";C

RUN
    X = 4.5, (X + 2) = 6.5, X/3 = 1.5
    and X squared = 20.25

If the value of X is re-assigned to X = 7, then a re-run of the latter version of the program results in:

    X = 7, (X + 2) = 9, X/3 = 2.33333333
    and X squared = 49

The output is still quite readable, but the value given for X/3 shows eight significant places after the decimal point. This is unnecessary and some stricter method of controlling the numerical output format would obviously be beneficial.

9.2 FORMATTED OUTPUT IN BASIC

In BBC BASIC, formatted output is achieved using the 'special' variable @%. The default field width is again ten characters, but this can be altered by assigning a particular value to @%. For example:

    @% = &20308

Note, first of all, that the number is in hexadecimal form since it is preceded with an ampersand.

The first number, 2 in the example, defines the form of the numerical output. A '2' gives the numerical output in decimal point form. A '1' gives the output in exponent form and a '0' gives the output in 'integer' form.

The next two numbers, 03 in the example, define the number of decimal places displayed. The last two numbers, 08, defines the total field width. If the required field width was ten, then the last two numbers would be 0A. For example:

```
10 @% = &20306
20 PRINT " PI = ";PI
```

```
RUN
PI = 3.142
```

For tabulated data, it may be necessary to change the output format during a printing list to accommodate the range of the numbers printed. Say the variables K, A(K), B(K) and C(K) are to be output in a FOR–NEXT loop in, for example, the general form:

```
***   *.**   ***.*   *.***E***
550   1.23   456.7   1.234E−10
```

Considering the field width to be set at say ten characters, the appropriate coding is:

```
@% = &0000A:P.K;: @% = &2020A:P.A(K);
: @% = &2010A:P.B(K);: @% = &1030A:P.C(K)
```

Precise alignment under column headings may thus be obtained. Note that the short form 'P.' has been used in place of the full command, PRINT. The coding might, at first glance, look very complex but all that is being done is to alter the output format specification to suit each variable just before it is actually printed.

Use of the @% variable to format numerical output is peculiar to BBC BASIC. For the IBM-PC using BASICA, or a compatible machine with a clone of BASICA, the output format may be specified with the command PRINT USING. For example:

```
10 PI = 3.141593
20 PRINT USING " #.###";PI
```

```
RUN
3.142
```

The PRINT USING statement may also be used to re-specify the output format. For the previous list of variables, i.e. K, A(K), B(K) and C(K), the appropriate print format coding in BASICA is:

PRINT USING "### #.### ###.# ##.###^^^^";K;A(K);B(K);C(K)

The format is specified using the hash symbol, #, and the spacing between the numbers is also defined within the quotation marks. The inclusion of four carets, ^, after the format specification indicates that the number is to be written in exponent form.

An alternative method of controlling the location of printed output is to use the TAB command. For example:

```
10 PRINT "012345678901234567890"
20 Z = 12
30 PRINT TAB(10);Z;TAB(15);2*Z
40 Z = Z + 1
50 IF Z > 13 GOTO 70
60 GOTO 30
70 END
RUN
 0 1 2 3 4 5 6 7 8 9 0 1 2 3 4 5 6 7 8 9 0
                     1 2         2 4
                     1 3         2 6
```

TAB(10) prints the value of Z ten spaces to the right. TAB(15) then prints the value of 2*Z fifteen spaces to the right.

For BBC machines, the TAB statement may also be specified in two dimensions. In teletext mode, i.e. MODE 7, the text coordinates are as indicated in figure 9.1.

Figure 9.1 BBC microcomputer, MODE 7 text coordinates

The use of PRINT TAB(x,y) allows text or data to be printed in any specified character 'cell' in the appropriate MODE. For example:

```
5 MODE 7:CLS
10 FOR K = 1 TO 1000
20 PRINT TAB(10,12)"COUNT = ";K
30 NEXT K
```

The two numbers, in brackets, after TAB specify the x and y coordinates where printing is to start.

If the graphics MODE 5 is to be used, there are 20 cells across the screen and 32 cells down. This resolution may be too coarse in some cases and it is useful to be able to position characters on a much finer grid. The statement VDU 5 enables text to be printed at the exact position of the graphics cursor. The statement MOVE, see section 9.3, may then be used to position the text. The graphics screen coordinates are indicated in figure 9.2.

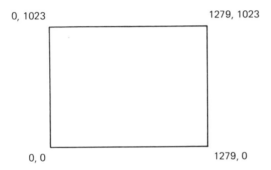

Figure 9.2 *BBC microcomputer, graphics screen coordinates*

```
10 MODE 4
20 VDU 5
30 MOVE 10,500
40 PRINT "Water is denoted by H O"
50 MOVE 682,484
60 PRINT "2"
70 VDU 4
```

The command VDU 4 separates the graphics and text cursors. On running the above program, the monitor will display:

Water is denoted by H_2O

For the IBM-PC and compatibles, the two-dimensional TAB statement is not available. These machines use the LOCATE statement as an exactly equivalent alternative. The IBM-PC text coordinates are as shown in figure 9.3.

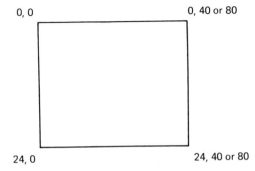

Figure 9.3   *IBM-PC text coordinates*

The horizontal resolution may be either 40 or 80 characters in length. It may be altered with the command WIDTH size, where the variable 'size' is 40 or 80 as appropriate. For example:

```
 5 CLS
10 WIDTH 40
20 FOR K = 1 TO 1000
30 LOCATE 12,20
40 PRINT "COUNT = ";K
50 NEXT K
```

9.3   COMPUTER GRAPHICS

Experimental data usually lends itself well to presentation in graphical form and the old adage, 'every picture speaks a thousand words', is particularly relevant in the context of physical systems. Much enlightenment can be gained from a single pictorial representation of some complex system or phenomenon, which might otherwise defy a lucid verbal description.

The production of neat and well balanced graphical information may be quite tedious and painstaking if done by hand, but this is an area where the modern microcomputer comes into its own. With a little experience, the generation of graphical output on a microcomputer becomes routine. Permanent 'hard copies' may be obtained using one of the many 'screen-dump' programs which are available and with a little ingenuity, a certain degree of animation may be incorporated into the monitor display. No other media quite offers this level of flexibility in the possibilities for presentation of data.

(i) Graphics on the BBC Microcomputer

In general, 'graphics' and 'text' are treated separately. The BBC microcomputer
MODE 7 text screen, figure 9.1, consists of 24 rows and 39 columns. The datum
point, (0,0), is located at the top left-hand corner of the screen. The graphics
screen, figure 9.2, has its datum at the bottom left-hand corner of the screen. A
summary of the MODEs available, resolution, colours and screen memory is
given in table 9.1.

Table 9.1   BBC microcomputer, summary of MODEs

Mode	Colours	Graphics	Memory	Text
0	2	640 × 256	20k	80 × 32
1	4	320 × 256	20k	40 × 32
2	16	160 × 256	16k	20 × 32
3	2	Text only	16k	80 × 25
4	2	320 × 256	10k	40 × 32
5	4	150 × 256	10k	20 × 32
6	2	Text only	8k	40 × 25
7	16	Tele-text	1k	40 × 25

Colour may be varied and the foreground logical colours are given in table 9.2.
The background colour is obtained by adding 128.

Table 9.2   BBC microcomputer, foreground logical colours

0	black	8	flashing black–white
1	red	9	flashing red–cyan
2	green	10	flashing green–magenta
3	yellow	11	flashing yellow–blue
4	blue	12	flashing blue–yellow
5	magenta	13	flashing magenta–green
6	cyan	14	flashing cyan–red
7	white	15	flashing white–black

The numerical codes given in table 9.2 apply in MODE 2. The MODE 1 and MODE 5 codes are:

0 black
1 red
2 yellow
3 white

The logical colours, as given, may be changed however. In say MODE 1, the colour blue (logical colour 4) is to be used in preference to the default logical colour black. The command VDU19 is used to implement the change. For example:

VDU19,a,b,0,0,0

where 'a' is the logical colour to be changed and 'b' is the actual colour required. For example:

VDU19,0,4,0,0,0

The above command will change black to blue.

The additonal command COLOUR 1, will produce red text and GCOL 0, 3 will produce white graphics. For example:

To produce red text on a blue background in MODE 1 graphics

10 MODE 1
15 REM black default background is changed to blue
20VDU19,0,4,0,0,0
25 REM red is selected for foreground
30 COLOUR 1
40 PRINT "RED ON BLUE"

To produce white line graphics, the following coding is added

50 GCOL 0,3
60 MOVE 10,10
70 DRAW 400,400

On running the above program, the text appears at the top left-hand corner of the screen and a white line is drawn from the bottom left-hand corner.

The main graphics commands in BBC BASIC are summarised below:

GCOL 0,m   sets colour for graphics with 'm' specifying the
            colour. (m + 128) sets the background colour

MOVE a,b   moves graphics cursor to the point (a,b)

DRAW x,y   draws line from the current cursor position to the
            point (x, y)

PLOT69,x,y plots a point at position (x,y)

Note: a point is rather small but a '*' can be plotted in the correct position by making allowance for the character pixels, (32 × 24), relative to the top left-hand corner in MODE 1. That is:

MOVE X − 16, Y + 12
PRINT "*"

VDU 5   combines the graphics and text cursors. This would
               be necessary immediately prior to executing the above
               two commands

VDU 4   separates the graphics and text cursors

A number of examples are included elsewhere in this book which illustrate the use of some of the above basic commands. See program listings given in sections 2.5 and 10.1.

As a general procedure for graph plotting, the following method might be used. The axes are say X, from zero to some convenient XMAX (0–1100), and the corresponding Y to YMAX (0–700).

The basic framework for the plot is shown in figure 9.4.

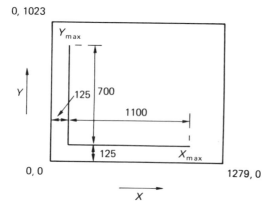

Figure 9.4 *Graph plotting reference frame*

The axes are plotted in the program with the instructions:

MOVE 125,825:DRAW 125,125:DRAW 1225,125

If the x and y values are calculated within the program, then the corresponding scaled values required for MOVE, DRAW or PLOT instructions are:

$X\% = 125 + INT(1100*(x/XMAX))$
$Y\% = 125 + INT(700*(y/YMAX))$

Although it is unneccessary to specify the X and Y values as integer quantities, it is more correct to do so.

Up to this point the entire screen has been used for text and graphics combined. In some cases, a more attractive display can be obtained by defining separate graphics and/or text windows.

A text window is defined as shown in figure 9.5.

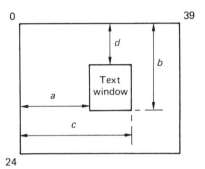

Figure 9.5   *Text window specification*

The text window is set up with the command:

VDU28,a,b,c,d

where a, b, c and d define the screen locations of the limits of the window in accordance with figure 9.5.

A graphics window is defined in much the same way, figure 9.6.

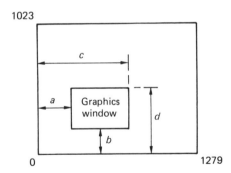

Figure 9.6   *Graphics window specification*

The graphics window is set up with the command:

VDU24,a;b;c;d;

Note the comma after 24 and the semi-colon after all the other values. Again a, b, c and d define the screen location of the graphics window in accordance with figure 9.6. For example:

```
10 MODE 1
20 REM Set up text window
30 VDU28,5,31,19,0
40 REM Set up graphics window
50 VDU24,800;200;1000;600;
60 REM Select red background for text window
70 COLOUR 129:CLS
80 REM Select yellow background for graphics window
90 GCOL 0,130:CLG
100 PRINT"WHITE ON RED"
110 REM Select black lines for graphics
120 GCOL 0,0
130 MOVE 500,400:DRAW 1200,500
140 END
```

On running the above program, the two windows are clearly depicted. The line drawn in the graphics window starts and ends outside the limits defining the window, but only that part of the line crossing the window is seen. Listing the program will show that all text is now confined within the text window. Line 10 of the program may be changed to MODE 5 and the program re-run to observe the effect.

As a further example, the program which follows sets up two graphics windows and two text windows, in various colours, as an indication of how the graphical output may be made more interesting. The sine function is plotted in black lines against a yellow background, while the cosine function is plotted as white points against a red background. The two text windows are similar, with white text on red and black text on yellow. Figure 9.7 shows the monitor display.

```
10 MODE 5
30 VDU24,100;600;600;1000;
40 GCOL0,138:CLG
50 GCOL0,0
60 MOVE 150,800:DRAW 550,800
70 MOVE 150,650:DRAW 150,950
80 FOR K=0 TO 50
90 DRAW 150+K*8,800+SIN(RAD(K*10))*100
100 NEXT
110 VDU24,700;600;1200;1000;
120 GCOL0,137
130 CLG
135 GCOL 0,7
140 MOVE 750,800:DRAW 1150,800
150 MOVE 750,650:DRAW 750,950
160 FOR K=0 TO 50
170 PLOT69,750+K*8,800+COS(RAD(K*10))*100
180 NEXT
190 VDU28,1,23,19,21
200 COLOUR 129
210 CLS
220 PRINT:PRINT"  SINE AND CCSINE"
230 VDU28,1,31,19,29
235 GCOL 0,138:CLS
240 COLOUR0:CLS
250 PRINT:PRINT" COMMAND ? ";
260 END
```

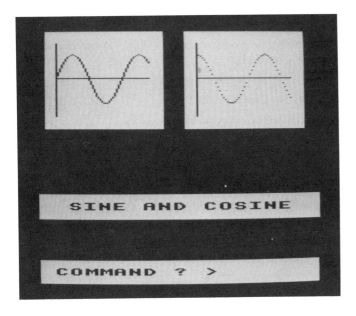

Figure 9.7 Monitor display using graphics and text windows

On changing line 10 of the above program to MODE 2, an interesting effect may be observed in the right-hand graphics window, where the cosine function is displayed. As an exercise, students should investigate how to produce the same, or a similar effect, in the left-hand graphics window.

The teletext mode on BBC machines does not support graphics, but all sixteen colours are available and various codes can be added to output information in colour to the monitor. The basic instruction is CHR$(n), with n as follows:

128 black	132 blue
129 red	133 magenta
130 green	134 cyan
131 yellow	135 white

For example, try:

PRINT "ABC";CHR$(129);"DEF";CHR$(130);"GHI"

To make the screen text flash on and off, use CHR$(136). For example:

PRINT CHR$(130);CHR$(136);"FLASH"

CHR$(137) turns the flashing characters off.

To produce a background colour other than black, three control codes are used with CHR$(157). For example:

PRINT CHR$(131);CHR$(157);CHR$(132);"Blue on Yellow background"

Double height characters may be produced using CHR$(141), but the instructions must be written twice. For example:

PRINT CHR$(141);CHR$(134);CHR$(157);CHR$(132);"DOUBLE HEIGHT"
PRINT CHR$(141);CHR$(134);CHR$(157);CHR$(132);"DOUBLE HEIGHT"

The above codes will produce large blue characters against a cyan background.

(ii) Graphics on the IBM-PC and Compatibles

The graphics capabilities of an IBM-PC depends on the expansion cards which are fitted to the machine and the type of monitor available. The usual extension hardware includes:

CGA   Colour Graphics Adaptor. Original IBM-PC graphics standard. Very low resolution and only a few colours

EGA   Enhanced Graphics Adaptor. Second generation IBM-AT graphics system. More colours and higher resolution than CGA

VGA   Video Graphics Adaptor. Current IBM graphics standard. More colours and higher resolution than EGA but only works with analogue monitors

SCD   Standard Colour Display

ECD   Enhanced Colour Display

MDPA   Monochrome Display/Printer Adaptor

The mode for either text or graphics is set with the SCREEN command. SCREEN 0 gives a text screen of 24 rows and 80 columns. The width of the screen may be altered with the command WIDTH 40. A summary of the available modes is given in table 9.3. Note that modes 7 to 10 inclusive require the Enhanced Graphics Adaptor.

Table 9.3 SCREEN modes for the IBM-PC and compatibles

0	text mode at current width (40 or 80 characters)
1	medium resolution graphics (320 × 200) and 4 colours
2	high resolution graphics (640 × 200), monochrome only
7	medium resolution graphics (320 × 200) and 16 colours
8	high resolution graphics (640 × 200) and 16 colours
9	enhanced high resolution (640 × 350), requires an enhanced colour display
10	monochrome high resolution graphics (640 × 350)

Colours for foreground, background and border are selected with the COLOR statement. Note the US spelling. The default settings are as given in table 9.4.

Table 9.4 Default colour settings for IBM-PC

0	black	8	grey
1	blue	9	light blue
2	green	10	light green
3	cyan	11	light cyan
4	red	12	light red
5	magenta	13	light magenta
6	brown	14	yellow
7	white	15	high intensity white

The colours 8 to 15 may be regarded as 'light', or 'high intensity' values for the corresponding colours 0 to 7.

In text mode, SCREEN 0, the screen colours are set with:

COLOR foreground,background,border

The variable 'background' must be in the range 0-7, 'border' can range between 0 and 15 and 'foreground' can range between 0 and 31. Values for 'foreground' in the range 16-31 produce flashing versions of the corresponding default colours in the range 0-15. For example:

COLOR 14,1,2

In SCREEN 0, the above command will set yellow text on a blue background with a green border.

The border option is not available in any of the graphics modes and the COLOR command is not supported in SCREEN 2 mode.

In all of the graphics modes, lines are drawn relative to the top left-hand corner of the screen, (0,0). The basic command is:

LINE (x1,y1)-(x2,y2)

A typical graphical framework, in medium resolution SCREEN 1, is illustrated in figure 9.8.

To plot x, y values evaluated within a program, the required scaling is:

X% = (250*x/XMAX) + 40
Y% = 170-(150*y/YMAX)

In the program which follows, the sine and square root functions are plotted in white lines against a blue background. The two colours are set with the

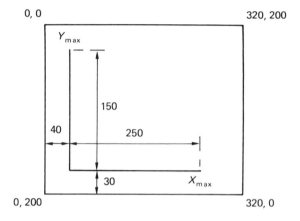

Figure 9.8 *Graphical framework in SCREEN 1 mode*

command COLOR a,b. The variable 'a' defines the background colour and 'b' defines the 'palette'. The background colour may range between 0 and 15, but palette can only enable three pre-determined colours. The pre-determined colours are listed in table 9.5.

Table 9.5   Pre-determined palette colours

Colour	Palette 0	Palette 1
1	green	cyan
2	red	magenta
3	brown	white

If 'palette' is an even number, then palette 0 is selected and the colours green, red and brown are associated with colour attributes 1, 2 and 3. If 'palette' is an odd number, then palette 1 is selected. Palette 1 is the default selection.

If the machine is fitted with an Enhanced Graphics Adaptor, the PALETTE command allows the user to define the preferred colours that they may wish to work with. For example

PALETTE 3,6

assigns the colour 6 to attribute 3. With a Standard Colour Display, colour 6 is brown. However, if an Enhanced Colour Display is available the effect may not be as expected and colour 6 appears as a shade of green. Various coloured lines are obtained by extending a graphics command, like LINE, as follows:

LINE (x1,y1)–(x2,y2),c

where 'c' is 1, 2 or 3 depending on the colour attribute required. If 'c' is left unspecified, the attribute defaults to the highest available in the palette, i.e. colour attribute 3.

```
10 CLS
20 SCREEN 1
30 COLOR 1,1:REM select background blue and palette 1
40 LINE (30,120)-(130,120)
50 LINE (30,120)-(30,20)
60 LINE (30,20)-(130,20)
70 LINE (130,20)-(130,120)
80 LINE (40,70)-(120,70)
90 LINE (40,110)-(40,30)
100 FOR K=0 TO 150
110 LINE (K/2+40,70-SIN(K/10)*30)-((K+1)/2+40,70-SIN((K+1)/10)*30)
120 NEXT K
130 LOCATE 17,5
140 PRINT "sine function"
150 LINE (170,120)-(270,120)
160 LINE (170,120)-(170,20)
170 LINE (170,20)-(270,20)
180 LINE (270,20)-(270,120)
190 LINE (180,110)-(260,110)
200 LINE (180,110)-(180,30)
210 FOR K=0 TO 800
220 LINE (K/10+180,110-SQR(K)*2)-((K+1)/10+180 ,110-SQR(K+1)*2)
230 NEXT K
240 LOCATE 17,23
250 PRINT "square root"
260 LOCATE 19,24
270 PRINT "function"
280 END
```

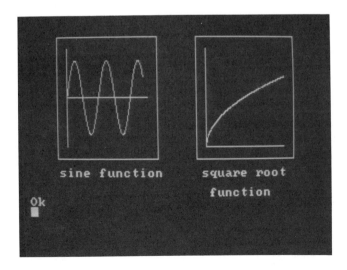

Figure 9.9 *Graphics output in SCREEN 1 mode*

The hard copy of the display, figure 9.9, is obtained by pressing the key marked 'PTR SC'. As a prerequisite, the graphics adaptor must first of all have been initiated from DOS with the command GRAPHICS. This is normally done before BASICA is entered. On IBM machines and compatibles, there is a further restriction in that a screen dump to a standard printer can normally be obtained in only the graphics modes SCREEN 1 and SCREEN 2.

Single points may be included in the plot with the command $PSET(x,y)$, where x and y describe the point coordinates.

Graphics windows may be defined with the command VIEW(x1,y1)–(x2,y2) where x1,y1 and x2,y2 correspond to the coordinates illustrated in figure 9.10.

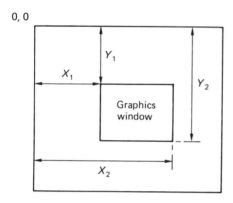

Figure 9.10 *IBM-PC graphics window*

The VIEW command may be further extended as follows:

VIEW (x1,y1)–(x2,y2),a,b

The variable 'a' allows the viewport to be filled in with colour. If 'a' is omitted then the viewport is not filled. The optional variable 'b' allows a boundary line to be drawn round the viewport. Both 'a' and 'b' are confined within the range defined by the colour palette.

When a viewport is specified, all subsequent graphics commands, e.g. LINE, are drawn with the coordinates relative to the top left-hand corner of the viewport.

In the program below, four separate graphics windows are defined with various mathematical functions, in various colours, drawn in each.

```
10 SCREEN 1:VIEW:CLS:KEY OFF
20 COLOR 6:REM select background brown and default palette 1
30 REM define graphics window with cyan border
40 VIEW (1,1)-(150,90),,1
50 LOCATE 2,2:PRINT "graphics window 1"
60 LINE (20,80)-(130,80)
```

```
70 LINE (20,80)-(20,20)
80 LOCATE 4,6:PRINT "y = x/2"
90 FOR K=0 TO 100
100 LINE (20+K,80-K/2)-(20+K+1,80-(K+1)/2)
110 NEXT K
120 REM define cyan graphics window with magenta border
130 VIEW (165,1)-(315,90),1,2
140 LOCATE 2,23:PRINT "graphics window 2"
150 LOCATE 4,27:PRINT " y = x"
160 LOCATE 3,27:PRINT "      2"
170 LINE (20,80)-(130,80),2
180 LINE (20,80)-(20,20),2
190 FOR K=0 TO 100
200 LINE (20+K,80-(K*K)/200)-(20+K+1,80-((K+1)*(K+1))/200),2
210 NEXT K
220 REM define cyan window with white border
230 VIEW (1,105)-(150,190),1,3
240 LOCATE 15,2:PRINT "graphics window 3"
250 LINE (20,80)-(130,80)
260 LINE (20,80)-(20,20)
270 CIRCLE (70,50),30,2:REM draw circle in magenta
280 REM define graphics window with cyan border
290 VIEW (165,105)-(315,190),,1
300 LOCATE 15,23:PRINT "graphics window 4"
310 LOCATE 17,23:PRINT "   y = x "
320 LOCATE 16,23:PRINT "        3 "
330 LINE (20,80)-(130,80),2
340 LINE (20,80)-(20,20),2
350 FOR K=0 TO 30
360 PSET (20+K*3,80-K*K*K/500)
370 NEXT K
380 LOCATE 23,1
390 END
```

The monitor display is shown in figure 9.11.

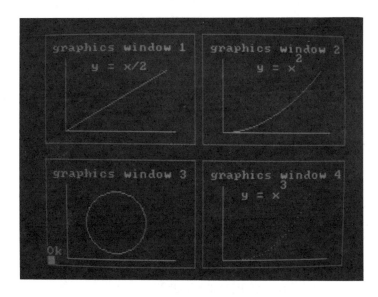

Figure 9.11 *Graphics windows on the IBM-PC*

In the program given, advantage is taken of the extended LINE command to draw some of the lines in either white or magenta. Note that this facility also extends to the CIRCLE and PSET commands.

In this section we have covered the basic commands and procedures used to produce a variety of colour graphics output. The material covered, although adequate is certainly not exhaustive and many more interesting effects can be devised using the extensive range of additional commands available in both BBC BASIC and IBM's BASICA.

9.4 HARD COPY OUTPUT, PRINTER CONTROL CODES

In the previous section the screen dumping of a graphics display to a printer has been periodically mentioned. The normal printed output, however, usually consists of tabulated numerical data and some variety can also be incorporated here to enhance the overall appearance to an acceptably professional level.

For BBC-based machines, the common printer control codes are:

VDU 2	switch printer on
VDU 3	switch printer off
VDU 1,14	enlarged characters on, cancelled by a carriage return
VDU 1,20	enlarged characters off
VDU 1,15	condensed characters on
VDU 1,18	condensed characters off
VDU 1,27,1,69	emphasised characters on
VDU 1,27,1,70	emphasised characters off

For example:

```
10 VDU 2
20 PRINT"         NORMAL PRINT"
30 PRINT
40 VDU 1,14
50 PRINT"   ENLARGED CHARACTERS"
60 PRINT
70 VDU 1,15
80 PRINT"              CONDENSED CHARACTERS"
90 VDU 1,18
100 PRINT
110 VDU 1,27,1,69
120 PRINT"        EMPHASISED CHARACTERS"
130 VDU 1,27,1,70
140 PRINT
150 PRINT"      NORMAL PRINT AGAIN"
160 VDU 3

RUN
```

NORMAL PRINT

ENLARGED   CHARACTERS

CONDENSED CHARACTERS

EMPHASISED CHARACTERS

NORMAL PRINT AGAIN

On the IBM-PC, the exact same output is obtained with the program below.

```
10 LPRINT"      NORMAL PRINT"
20 LPRINT
30 LPRINT CHR$(14);" ENLARGED CHARACTERS"
40 LPRINT
50 LPRINT CHR$(15);"   CONDENSED CHARACTERS"
60 LPRINT
70 LPRINT CHR$(27)"E"
80 LPRINT"     EMPHASISED CHARACTERS"
90 LPRINT CHR$(27)"F"
100 LPRINT
110 LPRINT CHR$(18)" NORMAL PRINT AGAIN"
RUN
```

NORMAL PRINT

ENLARGED   CHARACTERS

CONDENSED CHARACTERS

EMPHASISED CHARACTERS

NORMAL PRINT AGAIN

Note that on the IBM-PC, there are no separate commands to switch the printer on and off. Output to the printer is obtained with LPRINT. Similarly, a hard copy listing of a program may be obtained with LLIST.

9.5 DISC FILE HANDLING, DATA STORAGE AND RETRIEVAL

The storage of data on magnetic disc is one of the major advantages a microcomputer system has to offer its user. This facility may be utilised, for example,
to download data to a disc file during the execution of a particular program.
Another entirely different program may then be run in which the data stored on
the disc is retrieved and further processing performed. This could be particularly
convenient if either, or both, of the programs require large amounts of computer
memory. Alternatively, the disc file handling capabilities may be used, in the
most fundamental of roles, as a simple means of data storage and record keeping.

The more common file handling commands in BBC BASIC are summarised
below:

OPENIN	opens a file so that it may be read
OPENOUT	opens a new, or empty, file so that it may be written to
INPUT#	reads data from a file into memory
PRINT#	writes data to a file
CLOSE#	closes a file

For example:

```
10 INPUT" Today's date, eg. 21/08/89 - ";A$
20 X=OPENOUT "TEST DATA"
30 PRINT#X,A$
40 FOR K=1 TO 10
50 Z=K*K
60 PRINT#X,K,Z
70 NEXT K
80 CLOSE#X
90 END
```

The filename, 'TEST DATA' in this case, should not contain more than ten
characters. The program to read the data back from the file is quite straightforward. For example:

```
10 Y=OPENIN "TEST DATA"
20 INPUT#Y,A$
30 FOR I=1 TO 10
40 INPUT#Y,P,Q
50 PRINT" ";P;" ";Q
60 NEXT I
70 PRINT:PRINT" Date = ";A$
80 CLOSE#Y
90 END
```

The print statements in the above program are optional, but serve to check, in
the example, that the data has been read correctly. The only point to bear in
mind is that the data must be read in the exact same sequence that it was written
to the file.

IBM's file handling capabilities in BASICA operate in much the same manner although there are some significant differences. The main file handling commands in BASICA are summarised below:

OPEN   allows for input, or output, to a file
INPUT#   reads data from a file into memory
WRITE#   writes data to a file
CLOSE#   closes a file

The OPEN command is written generally as follows:

OPEN "character", #integer, "file-spec"

The variable 'character' may be specified as shown below:

O   specifies sequential output mode
I   specifies sequential input mode
A   specifies sequential append mode
R   specifies random input/output mode

The variable 'integer' defines a file number which must be referenced in all subsequent INPUT# or WRITE# commands.

The string 'file-spec' defines a file name which may also include the disc drive on which the file is to be saved to, or accessed from. If the drive is left unspecified, the file will be stored on drive C, the hard disc.

The command PRINT# is also available, although it is more cumbersome to use since data delimiters, i.e. commas, must be included in the PRINT# statement. Note that WRITE# does this automatically.

The following program in BASICA will store the given data in a file called data.dat on the floppy disc. For example:

```
10 INPUT" enter today's date - ";A$
20 OPEN"O",#1,"A:data.dat"
30 WRITE#1,A$
40 FOR K=1 TO 10
50 Z=K*K*K
60 WRITE#1,K,Z
70 NEXT K
80 CLOSE#1
90 END
```

Alternatively, using the PRINT# command, lines 30 and 60 of the above program would be altered to:

30 PRINT#1,A$
60 PRINT#1,K;",";Z

To retrieve the data from the floppy disc:

```
10 OPEN"I",#1,"A:data.dat"
20 INPUT#1,A$
30 FOR J=1 TO 10
40 INPUT#1,K,Z
50 PRINT"   ";K;"    ";Z
60 NEXT J
70 PRINT:PRINT"              ";A$
80 CLOSE#1
90 END
```

Since the screen display in a graphics, or text, mode is stored as an area of memory, it may also be saved as a disc file. This feature may be usefully employed to capture screen images for records and subsequent retrieval.

On BBC-based machines, the procedures to save a screen display are as follows.

Assuming that the current screen in MODE 1 shows the required picture, the display may be saved with:

*SAVE PICTURE 3000 + 4FFF

where 'PICTURE' is the chosen file name.

If the bottom two lines of the display are not required, then 4B00 should be substituted for 4FFF.

To retrieve the file at some later stage and re-display the screen, the following program is run:

10 CLS
20 MODE 1
30 *LOAD PICTURE

In the tele-text MODE 7, much less memory is required for the screen display and it may be saved with the command *SAVE PICTURE 7C00 + 3FF. Similarly, to omit the bottom two lines, 3E7 would be substituted for 3FF.

On IBM-PCs or compatibles, screen displays may be saved in a similar manner. The commands and procedures used with IBM type machines are as follows.

In SCREEN modes 1 and 2, the 16K screen buffer for CGA or EGA is set at &HB8000. The DEF SEG statement must therefore be used, first of all, to set up the segment address to the start of the screen buffer. To save the current screen display, the following coding is used:

210 DEF SEG = &HB800
220 BSAVE "A:PICTURE",0,&H4000

In line 220, the command BSAVE will save the display to a file named 'PICTURE' on the floppy disc. The offset of 0 and the length of &H4000 specifies that the entire 16K screen buffer is to be saved. If A: is omitted in line 220, the file will be saved to the hard disc.

If a monochrome display and parallel printer adaptor is used, the segment address of the 4K screen buffer is &HB0000. This should be substituted for &HB8000 in line 210 if SCREEN 0 is specified, or if a monochrome monitor is used.

Note that in the example programs, the trailing zero of the number assigned to DEF SEG is omitted. This is necessary because of the way the machine calculates actual memory addresses by adding the offset to the segment address. This is explained more fully in section 8.5.

To retrieve the file and re-display the screen, the following coding is used:

```
10 SCREEN 1
20 CLS
30 DEF SEG = &HB800
40 BLOAD "A:PICTURE",0
```

Again if A: is omitted, then 'PICTURE' will be loaded from the hard disc. Note: BSAVE and BLOAD cannot be used in any of the 640 × 350 resolution graphics modes.

9.6 A SIMPLE VIRTUAL INSTRUMENT

The virtual instrument concept was introduced in section 6.8. As an example of this graphics concept, the following program simulates a typical analogue type voltmeter in a practical application.

```
10 REM Virtual Instrument Demonstration
20 MODE 1
30 VDU19,0,4,0,0,0
40 MOVE 100,100:DRAW 1000,100
50 MOVE 100,100:DRAW 100,700:DRAW 360,700
60 MOVE 360,690:DRAW 360,710
70 MOVE 380,670:DRAW 380,720
80 MOVE 380,700:DRAW 640,700:DRAW 640,600
90 N=0
100 FOR K=1 TO 19
110 N=N+1
120 IF N/2<>N DIV 2 THEN Z=-1 ELSE Z=1
130 DRAW 640+20*Z,600-20*K
140 NEXT K
150 DRAW 640,200:DRAW 640,100
160 MOVE 1000,100:DRAW 1000,200
170 MOVE 1000,500:DRAW 1000,600
180 DRAW 730,600:DRAW 730,350
190 MOVE 670,350:DRAW 750,350
200 MOVE 670,350:DRAW 680,360
210 MOVE 670,350:DRAW 680,335
220 VDU 5
230 MOVE 280,780:PRINT"1.8 Volts"
240 VDU 4
250 VDU24,800;200;1200;500;
260 GCOL0,130:CLG
270 VDU24,810;210;1190;490;
280 GCOL0,137:CLG
290 VDU24,830;230;1170;470;
300 GCOL 0,131:CLG
310 GCOL 0,1
320 VDU 5
330 MOVE 920,300:PRINT"Volts"
340 MOVE 840,330:PRINT"0"
350 MOVE 1130,330:PRINT"2"
```

```
360 VDU 4
370 MOVE 850,450:DRAW 1150,450
380 DRAW 1150,350:DRAW 850,350:DRAW 850,450
390 MOVE 858,400:DRAW 850,430
400 MOVE 1142,400:DRAW 1150,430
410 MOVE 1000,420:DRAW 1000,450
420 MOVE 930,415:DRAW 925,440
430 MOVE 1070,415:DRAW 1075,440
440
450 *FX16,1
460 *FX190,8
470 REPEAT:UNTIL ADVAL(0)=256
480 VOLTS=ADVAL(1) DIV16*1.8/4095
490 PROCpoint(VOLTS)
500 GOTO 470
510
520 DEFPROCpoint(V)
530 R=440
540 ANG=18*V-18
550 GCOL 3,127
560 MOVE 1000+R*(SIN(RAD(ANG))),350
570 DRAW 1000+(R+30)*SIN(RAD(ANG)),410-20*SIN(RAD(ABS(ANG)))
580 GCOL 3,3
590 MOVE 1000+R*(SIN(RAD(ANG))),350
600 DRAW 1000+(R+30)*SIN(RAD(ANG)),410-20*SIN(RAD(ABS(ANG)))
610 ENDPROC
```

Using the reference voltage of the on-board A/D converter on the BBC micro-computer, a potentiometer may be connected across this voltage source and the output fed to pins 15 and 8 of the analogue input port, i.e. channel 1. This provides an infinitely variable voltage input ranging between 0 and 1.8 volts. The program given continuously samples the input channel and displays the results on an 'animated voltmeter' on the screen.

Figure 9.12 shows a screen display for a particular voltage input level. The 'live' situation is much more dramatic however, with the voltmeter's pointer actively responding to changes in the potentiometer setting.

As an exercise, this example could be programmed to run on an IBM-PC fitted with an appropriate A/D expansion card.

9.7 SYSTEM MONITORING

The phrase 'system monitoring' encompasses many of the aspects which form the substance of this book. The process of system monitoring includes the basic measuring sensors, signal conditioning, A/D conversion, the computer interface and the display of the results. On interpretation of the results, the control functions may then be brought into play and thereby close the control loop. From the point of view of an operator, the system monitor ultimately appears in the form of a graphics 'mimic' of the actual system, be it plant, process or procedure. The mimic relays all of the relevant status of the system and the operator may then invoke any number of actions based on the information at his or her disposal. Much of the control aspects would be programmed to occur automatically.

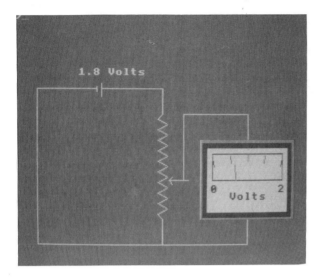

Figure 9.12 *A simple virtual instrument*

In section 6.6 a programming example is included on an application where a quantity of water is to be maintained at a pre-selected temperature. The water temperature is continuously monitored and various actions are implemented depending on the actual water temperature and its relation to the pre-selected value. If the actual temperature is below the selected value, then a heater is switched on while a cooling fan is switched off. If the actual temperature is above the selected value, then the heater is switched off while the fan is switched on. If the measured temperature lies within a narrow band of the selected value, then both the heater and fan are switched off. The tolerance band is prescribed by the user and forms part of the manual input data.

The actual hardware is shown in figure 9.13 and the status of the system can be conveniently displayed in graphics, see figure 9.14.

The display on the mimic is continuously updated and any changes which occur in the system, e.g. switching of fan or heater, are immediately indicated on the screen. A temperature–time record is also displayed and the use of graphics in this example shows how the physical system may be represented in clear, informative and easily understood terms. The temperature–time graph might be considered as a simple visual instrument chart recorder.

A second example, related to a more complex piece of mechanical hardware, is illustrated in figures 9.15, 9.16 and 9.17. Figure 9.15 shows the actual hardware, which consists of an air conditioning unit including a centrifugal fan, a number of electrical heaters, a steam injection system and an air chiller. The air condition at inlet and outlet is monitored with humidity sensors and thermo-

Figure 9.13 *Water temperature control system*

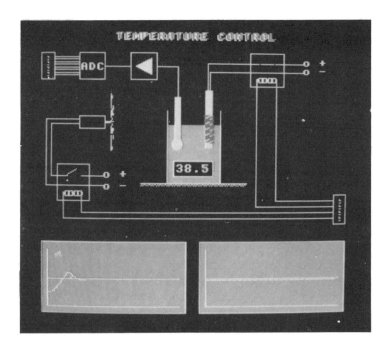

Figure 9.14 *Graphics mimic of water temperature control system*

Figure 9.15 *Air conditioning system*

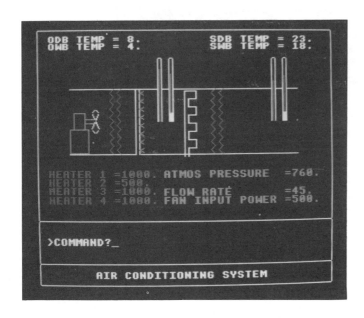

Figure 9.16 *Graphics mimic of air conditioning system*

couples, and the volumetric flow rate of the air at outlet is measured with a pressure differential flowmeter. Figure 9.16 shows the mimic of the system where the status of all the individual elements is monitored and indicated, in graphics, as either active or inactive. In figure 9.17 the thermodynamic processes undergone by the air are depicted on a typical psychrometric chart. Control signals may be sent out via the computer terminal to switch, or adjust, the system elements as required for a particular air condition at outlet.

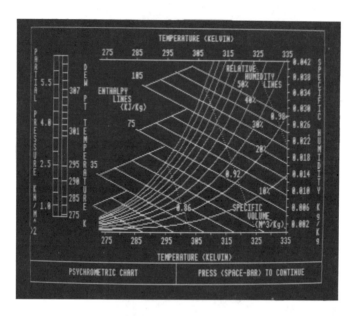

Figure 9.17 *Air conditioning process*

This particular example comes from the field of thermodynamics and it is not intended here to describe the relevant background theory. The example shows nonetheless, how a complex industrial process, and its control, may be effeccc-tively represented in computer graphics. In modern industrial plant and process industries, this form of representation is fairly common.

EXERCISES

1. Write a program to produce a set of mathematical tables including the follow-ing functions:

X	X^2	X^3	$\sqrt{X}$	$\exp(X)$	$\ln(X)$

The printed output should be neatly formatted and should include suitable headings as appropriate.

2. In a test on a centrifugal pump running at a constant speed N, the pressure rise Δp and input torque T are measured for various flow rates Q. This data is to be processed to give the pump efficiency E from:

$$E = 100\,[(Q\,\Delta p)/(\omega T)]$$
where $\omega = 2\pi N/60$

Write a program to evaluate the variation of E with Q using arrays to store each result.

The pump speed and the number of sets of readings taken, are to be input interactively at the keyboard.

All other data, i.e. Q, Δp and T, are to be read as a data block.

Arrange the print out of the processed data as given below:

CENTRIFUGAL PUMP CHARACTERISTICS

Test No.	Quantity (litres/s)	Efficiency (%)
##	##.##	##.##

Run the program using the following test data:

Pump speed = 1500 rev/min

Flow rate	– (litres/s) 0.0	10.5	20.0	30.0	40.5
Pressure difference –	(kN/m²) 182.2	190.2	191.2	183.3	152.9
Input torque	– (N m) 22.6	34.6	44.4	53.2	57.0

3. Set up three graphics windows according to the following approximate format:

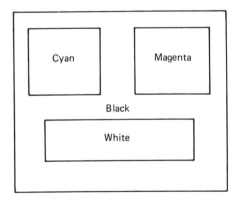

Draw a circle in one window, a square box in another and list your program in the third.

4. Develop a 'user friendly' interactive program, incorporating graphics, to demonstrate the phenomenon of damped vibration with light damping. Given:

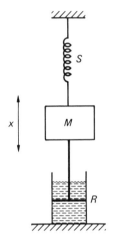

The equation of motion is

$$M\ddot{x} + R\dot{x} + Sx = 0$$

with the solution for light damping $(\xi < 1)$

$$x = A \exp(-\xi \, \omega_n \, t) \sin(\omega_d \, t + \phi)$$

where: $\xi = R/[2\sqrt{(S\,M)}]$
$\omega_n = \sqrt{(S/M)}$
$\omega_d = \omega_n \sqrt{(1 - \xi^2)}$
$\phi = \tan^{-1}\{(\omega_d \, x_0)/(\dot{x}_0 + \xi \, \omega_n \, x_0)\}$
$A = x_0/\sin \phi$

The period of the oscillation is: $2\pi/\omega_d$.
Input parameters: M, R, S, x_0 and $\dot{x}_0$.
Output: graph of x against time over the duration of four periods (a suitable time step length for the plot is period/50).
Apply to following data: $M = 1$ kg, $S = 100$ N/m, $R = 10$ N s/m,
$$x_0 = 0.2 \text{ m and } \dot{x}_0 = 0 \text{ m/s.}$$

5. If the system in 4 is subjected to a periodically varying force, $F_0 \sin(\omega t)$, where F_0 is the amplitude and ω the frequency, examine using graphics, the variation of amplitude with frequency for damping factors, ξ, in the range 0–1. Use dimensionless plots of (x_0/x_s) against (ω/ω_n).
Given:

$$\frac{x_0}{x_s} = \frac{1}{\sqrt{[(1 - y^2)^2 + 4\xi^2 \, y^2]}}$$

where $x_s = F_0/S$ and $y = \omega/\omega_n$.

Note that a resonant condition occurs at $\xi = 0$ and $\omega = \omega_n$, which results in infinite amplitudes of oscillation.

6. Using the simple virtual instrument described in section 9.6, re-write the graphics routine to simulate a digital type of voltmeter.
This exercise may also be written for an IBM-PC fitted with a suitable A/D expansion card.

REFERENCES

Acorn Ltd, *The BBC Microcomputer System, Master Series Guide*, Acorn Computers, 1986.

Borland Inc., *Turbo-Basic, Owner's Handbook*, Borland International, 1987.

Coll, J., *The BBC Microcomputer User Guide*, BBC, 1982.

Cownie, J., *Creative Graphics on the BBC Microcomputer*, Acornsoft, 1982.

Cryer, N., Cryer, P. and Cryer, A., *Graphics on the BBC Microcomputer*, Prentice-Hall, 1983.

IBM Corp. Ltd, *BASIC 3.20 Reference*, IBM, 1986.

McGregor, J. and Watt, A., *The Art of Graphics for the IBM-PC*, Addison-Wesley, 1986.

Chapter 10
Applications

In this final chapter it is the intention to provide some detailed descriptions of the application of the principles which have been covered in the preceding pages of the book. We believe that much can be learned from example, albeit passive example. We hope therefore that a description of the solutions devised for the problems considered will prove useful to those about to incorporate microcomputer technology into their measurement systems.

10.1 STUDIES OF A VIBRATING CANTILEVER

Mechanical vibrations are often of a fairly low frequency and for a large number of cases a relatively slow data-acquisition rate would normally suffice. To measure the vibratory response of a cantilever, some means of translating the deflection into output voltage is required. An obvious solution would be to incorporate a strain measuring sensor into the system and this may be done using a strain gauge bridge. Figure 10.1 depicts a typical physical system and indicates a suitable configuration of strain gauge elements in a resistive bridge circuit.

The cantilever may be made of any suitable elastic material. In the example given, aluminium bar of cross-section 25 mm by 6 mm was used.

Gauges 1 and 2 are aligned parallel to the cantilever and will respond, in equal and opposite manner, to any applied load which causes a deflection. Gauges 3 and 4 are aligned at 90 degrees to gauges 1 and 2 and they are insensitive to the loading state although they provide temperature compensation. The bridge output voltage could be doubled, if required, by aligning gauges 3 and 4 in the same direction as gauges 1 and 2.

Using the strain gauge amplifier described in section 2.3, a suitable voltage output may be generated for a microcomputer interface. A static load test on

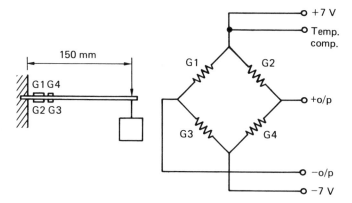

Figure 10.1 *Vibrating cantilever and sensor arrangement*

the cantilever should then result in a calibration similar to that shown in figure 10.2.

Using the analogue input terminal on the BBC microcomputer, see section 7.1, the amplifier output may be connected with the + o/p line to pin 15 and the − o/p line to pin 8. The amplifier zero adjustment potentiometer should be set such that an output of approximately 1 volt is generated for the null deflection condition. This will ensure that output voltages generated between the maximum and minimum deflection, for a step input, are contained within the range 0 to 1.8 volts.

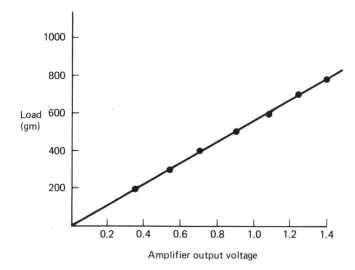

Figure 10.2 *Static calibration of loaded cantilever*

The program given below captures the data at about 240 Hz and then evaluates the signal frequency. An optional graphics output is included.

```
 10 @%=&20307
 20 *FX16,1
 30 *FX190,8
 40 MODE 7
 50 CLS
 60 DIM I(200),P(200),PL(200),PK(100),X(100)
 70 INPUT"GRAPHICS OUTPUT REQd Y/N ";Q$
 80 PRINT:PRINT
 90 IF Q$="N" GOTO 110
100 INPUT"GRAPHICS SCALE FACTOR (1 TO 50) ";SC
110 PRINT:PRINT"Computation in progress"
120 TIME=0
130 FOR K=1 TO 200
140 REPEAT:UNTIL ADVAL(0)=256
150 I(K)=ADVAL(1)
160 NEXT
170 T=TIME/100
180 FS=200/T
190 FOR K=1 TO 200
200 P(K)= ((I(K) DIV256)*1.8/255)-2.2*(-1)
210 NEXT
220 SUM=0
230 FOR K=1 TO 200
240 SUM=SUM+P(K)
250 NEXT
260 MEAN=SUM/200
270 FOR K=1 TO 200
280 PL(K)=P(K)-MEAN
290 NEXT
300 MAX=PL(1)
310 X(1)=1
320 FOR K=2 TO 200
330 IF PL(K)<MAX GOTO 360
340 MAX=PL(K)
350 X(1)=K
360 NEXT
370 PK(1)=MAX
380 N=2
390 FOR K=X(1)+1 TO 198
400 IF PL(K)<PL(K+1) AND PL(K+1)>=PL(K+2) THEN PK(N)=PL(K+1)
410 IF PL(K)<PL(K+1) AND PL(K+1)>=PL(K+2) THEN X(N)=K+1
420 IF PL(K)<PL(K+1) AND PL(K+1)>=PL(K+2) THEN N=N+1
430 NEXT
440 AL=0
450 FOR K=1 TO N-2
460 AL=AL+(X(K+1)-X(K))
470 NEXT
480 AL=AL/(N-2)
490 FR=(X(N-1)-X(1))/(T*AL)
500 PRINT:PRINT"Sampling frequency = ";FS;" Hz"
510 PRINT:PRINT"Signal frequency = ";FR;" Hz"
520 IF Q$="N" GOTO 830
530 MODE 4
540 MOVE 10,10
550 DRAW 10,800
560 MOVE 10,400
570 DRAW 1010,400
580 MOVE 10,400
590 FOR K=1 TO 200
600 DRAW K*SC+10,PL(K)*500+400
```

```
610 NEXT
620 PRINTTAB(5,1)"FREQUENCY RESPONSE OF "
630 PRINTTAB(5,3)"A VIBRATING CANTILEVER"
640 PRINTTAB(5,6)"SAMPLING FREQUENCY = ";FS;" Hz."
650 PRINTTAB(5,8)"SIGNAL FREQUENCY = ";FR;" Hz."
660 MOVE X(1)*SC+10,10
670 DRAW X(1)*SC+10,100
680 FOR K=1 TO N-1
690 MOVE X(K)*SC+10,0
700 DRAW X(K)*SC+10,50
710 NEXT
720 FOR K=0 TO N-2
730 MOVE X(1)*SC+10+K*AL*SC,100
740 DRAW X(1)*SC+10+K*AL*SC,110
750 NEXT
760 PRINTTAB(5,10)"REDRAW TO ANOTHER SCALE Y/N ";
770 INPUT A$
780 IF A$="N" GOTO 830
790 PRINTTAB(5,12)"GRAPHICS SCALE FACTOR (1 TO 50) ";
800 INPUT SC
810 CLS
820 GOTO 540
830 END
```

The program may be easily extended to calculate the amplitude logarithmic decrement and the damping factor. An informative study may also be made on the cantilever's response with added masses and/or variable damping included.

Figure 10.3 shows a typical graphics output with the cantilever very lightly damped. The horizontal scaling factor, SC, was 4. The longer vertical lines indi-

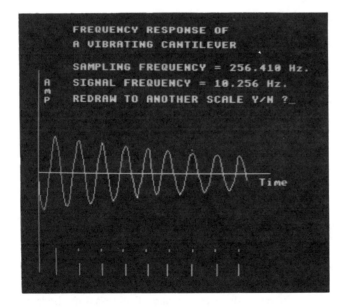

Figure 10.3 Graphics output for vibrating cantilever

cate the positions of the positive maxima in the signal identified by the processing algorithm. The shorter vertical lines indicate the locations of positive maxima according to the estimated signal frequency.

10.2 INTERNAL COMBUSTION ENGINE INDICATOR

In studies of internal combustion engine performance, the indicated power of the engine provides a useful measure of the efficiency of the combustion processes. The indicated power is evaluated from a record of the relationship between the pressure and volume within an engine cylinder during a working cycle. In fact the volume is not measured directly, but is inferred from a measurement of the crank angle, the two being simply related. Figure 10.4 shows a representative pressure–volume trace for a typical four-stroke petrol engine.

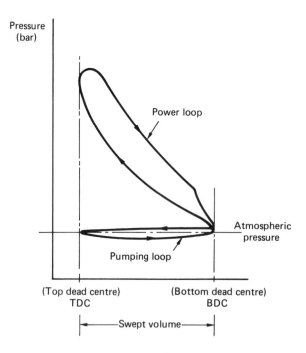

Figure 10.4 *Pressure–volume trace for a four-stroke engine*

The indicated power of a single cylinder engine is evaluated as the product of the area enclosed by the power loop, less the area enclosed by the pumping loop, and the engine speed in revolutions per second. Traditional methods of measuring indicated power were basically mechanical in nature. Recent developments however, based on microcomputer systems, have largely superseded the older mechnical devices. For a fuller exposition on methods of engine testing the reader should consult the standard thermodynamics texts given at the end of the chapter.

The basic requirement, then, is to monitor simultaneously the cylinder pressure and the crank angle. The cylinder pressure varies quite rapidly and common practice is to use a piezoelectric type of pressure transducer. Piezo-electric transducers have very fast response times and they are particularly well suited for dynamic pressure measurement. The measurement of crank angle offers more flexibility in the choice of transducer, but a magnetic pick-up, see section 2.2, may be perfectly adequate for the purpose. Figure 10.5 shows a schematic layout of the engine instrumentation and data-acquisition system.

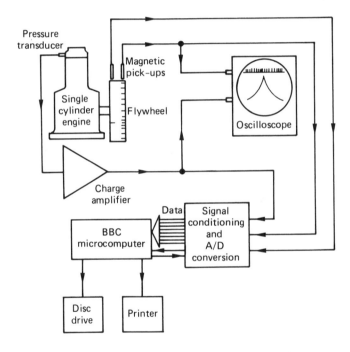

Figure 10.5 *Engine instrumentation and data-acquisition system*

The instrumentation includes two magnetic pick-ups. One is used to detect top dead centre, i.e. TDC, which is marked with a single separate groove machined in the periphery of the engine flywheel. The second magnetic pick-up is used to detect the passage of crank angle markers, small grooves in the flywheel, which are spaced at 10 degree intervals of rotation. At both top dead centre and bottom dead centre, there are additional grooves giving two extra markers on either side of top and bottom dead centre, with a 5 degree spacing. With this arrangement of crank angle marking, one revolution of the flywheel generates 40 pulses from the second magnetic pick-up. Two flywheel rotations, i.e. one engine cycle, generates 80 output pulses. Conversely, it is apparent that 240 output pulses encompass three complete and consecutive engine cycles. In the data-acquisition software which was developed for this application, the single top dead centre pulse was used as a prompt to start the data-acquisition routine. Subsequently, 240 consecutive pressure measurements were taken and stored using the crank angle marker pulses to generate a start conversion signal for the A/D converter. The pressure data was then averaged over the three cycles and an analysis performed on the averaged data.

In their basic forms, the transducer output voltages are unsuitable for direct interfacing to the A/D converter or to the microcomputer user port. Signal conditioning for the TDC and crank angle markers involved the use of a voltage comparator, LM311N, such as RS 308-843. The comparators were used to detect when a voltage signal reaches a set threshold value. The comparator outputs, which are TTL compatible, could then be interfaced to the start conversion pin of the A/D converter, or the CB2 line to the BBC microcomputer user port, see section 7.2. The signal conditioning circuit for the crank angle marker is shown in figure 10.6.

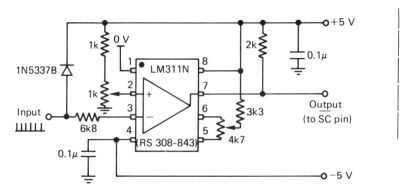

Figure 10 6 Crank angle marker signal conditioning

The comparator output is either 0 V or 5 V depending on the relative states of the inputs to pins 2 and 3. If the voltage at pin 2 is more positive than that at pin 3 then the comparator output is 5 V, or logic '1'. If the voltage at pin 3 becomes more positive than that at pin 2, then the comparator output is 0 V, or logic '0'. The input signal to pin 3 is a series of pulses generated from the magnetic pick-up monitoring the passage of crank angle markers. These pulses were greater than +5 V and a diode with a conducting voltage of 4.7 V was added to the circuit to protect the comparator from excessive voltage input to pin 3. The comparator threshold voltage was adjusted through the variable 1k resistor in the input line to pin 2. A suitable setting was obtained by trial and error. The 0.1 μF capacitors shown in the diagram were included to attenuate inherent noise in the circuit.

The effect of the passage of one input pulse is to change the comparator output from a logic '1' to a logic '0' and then back to logic '1' until the arrival of the next pulse. Referring back to figure 6.11, it is apparent that in this form, the comparator output may be used as a start conversion prompt for an A/D converter.

The external trigger, to start the data-acquisition cycle at TDC, was conditioned in much the same way. The extremely low level of the output voltage from this transducer, however, required a modified circuit. The layout is depicted in figure 10.7.

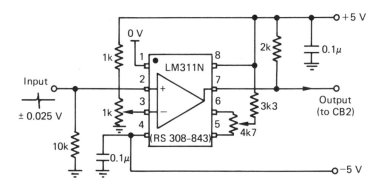

Figure 10.7 *Top dead centre trigger signal conditioning*

The 10k resistor on the input line to pin 2 induces a small current to flow through the resistor. Since the input impedance to pin 2 is massive, a large enough voltage, to switch the comparator, is induced at pin 2 on the passage of the triggering pulse. The threshold setting on pin 3 was again determined by trial and error.

Signal conditioning for the pressure transducer involved the use of an additional amplifier, LF355N, such as RS 307-058, shown schematically in figure 10.8.

The output from the pressure transducer charge amplifier was a negative voltage with a −2 V offset. The additional amplifier was therefore connected in the inverting mode to produce a positive only output for the A/D converter. A balance circuit was also included to zero the output at atmospheric pressure. In fact, the balance circuit was set such that a small positive output was generated at atmospheric pressure. This is a necessary feature since the engine cylinder pressure will fall slightly below the ambient pressure on a suction stroke, and a uni-polar input is required for the A/D converter.

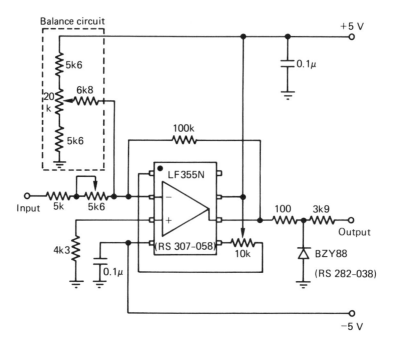

Figure 10.8 Pressure transducer, additional signal conditioning

The 5.6k variable resistor on the input line to pin 2 was used to adjust the amplifier output voltage. In normal operation, the resistor was set such that the peak cylinder pressure resulted in an output signal of approximately 3 V. The Zener diode in the output line had a conducting voltage of +3.3 V and was included to limit the output voltage and protect the relevant pin on the A/D converter.

Calibration of the pressure transducer and signal conditioner was performed by running the engine and then switching off the ignition while simultaneously opening the throttle valve to its fullest extent. This ensures that a full charge of air is drawn into the engine cylinder; compression then occurs but no firing takes place. With the bore, stroke and approximate compression index for the engine all known, the peak pressure could be evaluated and related to the maximum voltage generated. Since the voltage at atmospheric pressure was also known, then the linear calibration constant could be obtained.

The A/D converter used in the system, ZN427E-8, such as RS 309-464, is shown in figure 10.9.

The analogue voltage input from the pressure transducer conditioner was passed to pin 6 on the chip. Pin 4 takes the start conversion signals from the

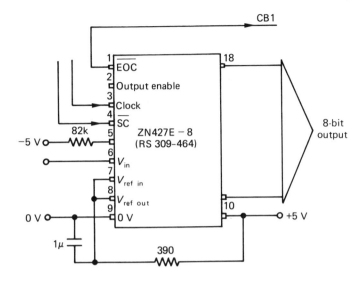

Figure 10.9 A/D converter for internal combustion engine

crank angle marker conditioning circuit and the end of conversion signal is routed from pin 1 to the CB1 line on the BBC microcomputer user port.

Also included is a necessary clock signal to time the operation of the A/D converter. The timing signal was provided by a programmable crystal oscillator, PXO600, such as RS 301-858, which was connected to provide a 600 kHz square-wave pulse train to pin 3 of the converter. Figure 10.10 shows the clock circuit diagram.

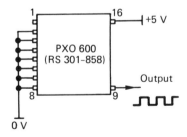

Figure 10 10 Clock circuit diagram for 600 kHz operation

The program which follows performs the basic data-acquisition function. The specific details of the analysis of the captured data and the graphics output are omitted for clarity.

```
 10 REM part of program to capture 80 pressure values
 20 REM for each of 3-cycles (6-revolutions) at known
 30 REM crank angle positions.
 40 REM the BASIC routine is included to average the
 50 REM pressure data over the 3 cycles.
 60
165 HIMEM=&3FFF
168 DIM PT(80)
170 DIM PICT 200
175 FOR Z%=0 TO 3 STEP 3
177 P%=PICT
180 [OPT Z%
190 LDX #&00            \set data counter to zero
200 LDA #&50            \set PCR to give CB1 and CB2 as input
210 STA &FE6C
220 LDA #&00            \set user port as input
230 STA &FE62
240 LDA &FE60           \clear IFR
250 .loop LDA &FE6D  \check for TDC on CB2
260 AND #&08
270 BEQ loop
280 .dat   LDA &FE6D  \check for EOC on CB1
290 AND #&10
300 BEQ dat
310 LDA &FE60           \read port
320 STA &4000,X         \store data from &4000
330 INX
340 CPX #&F0
350 BNE dat
360 RTS
370 ]
380 NEXT Z%
385 CALL PICT
390 REM AVERAGE AND STORE
395 DIM PT(79)
400 FOR I=0 TO 79
410 PT(I)=(?(&4000+I)+?(&4050+I)+?(&40A0+I))/3
420 ?(&4100+I)=PT(I)
430 NEXT I
440
450
```

The program continues with an analysis of the pressure–volume data. The areas enclosed by the power and pumping loops are evaluated and various engine performance parameters are calculated. Some care is required in matching the averaged pressures to the appropriate volumes but the particular details are too specific to be of general interest. A typical graphical output is given in figure 10.11.

The system as given is quite rudimentary and much more sophistication could be built into the data-acquisition routine, the calibration of the instrumentation and the analysis of the captured data. Many more complex systems are in existence but the basic features are very similar to those described for this application. Further details of alternative systems are available from the wealth of publications available in the literature, see references.

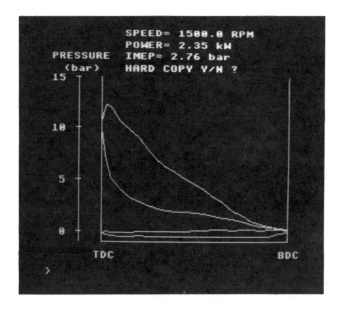

Figure 10.11   *Pressure–volume diagram for internal combustion engine*

10.3 A MICROCOMPUTER-BASED TEST FACILITY FOR BOILING HEAT TRANSFER STUDIES

Boiling heat transfer is a commonly occurring phenomenon in many engineering systems. The complexity of most practical systems denies a wholly analytical solution and experimental investigations are often aimed therefore at establishing the basic governing parameters in any given practical system. The experimental test facility described here was developed to investigate the critical heat input rates just to initiate localised nucleate boiling from a small laser flash lamp. The coolant consisted of a 60/40 mixture of ethylene glycol and water and the system variables included:

1. the coolant flow rate
2. the coolant mean bulk temperature
3. the energy input to the flash lamp.

   Figure 10.12 shows a schematic diagram of the physical system to be investigated:

   To vary and control the experimental variables a scaled-up test facility was built to incorporate the following essential features:

1. variable flow rate
2. a flow measurement device
3. control of the fluid mean bulk temperature

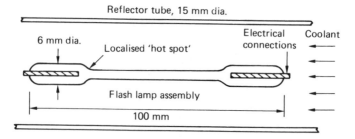

Figure 10.12 Flash lamp and reflector assembly

4. control of the heat input rates to the flash lamp assembly
5. the measurement of the salient temperatures
6. a microcomputer-based data-acquisition and control system.

   The experimental test facility is shown schematically in figure 10.13. Temperatures are measured at three important locations and the pressure is measured upstream and downstream of the flow monitoring device.

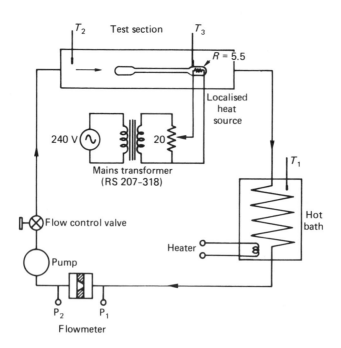

Figure 10.13 Experimental test facility

The localised heat source and the flow rate are adjusted manually. The hot bath temperature is monitored and controlled by the microcomputer-based system and all temperatures and the differential pressure across the flow meter are scanned in sequence during a test run. Visual judgement is used to determine the onset of localised nucleate boiling, defined as 'a regular formation of vapour bubbles in the vicinity of the localised heat source'.

(i) Flow Measurement

A simple orifice plate formed the basis of the flow measuring device. The physical dimensions were designed according to BS 1042 standard and the upstream and downstream pressures were sensed using inexpensive piezo-resistive pressure transducers, such as RS 303-343. The flowmeter signal conditioning circuit is illustrated in figure 10.14.

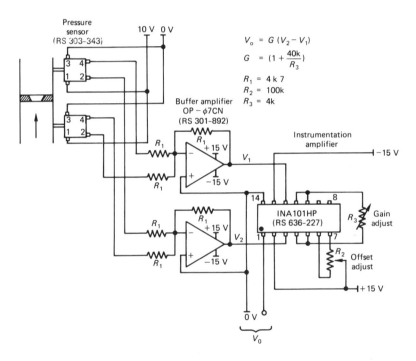

$$V_o = G (V_2 - V_1)$$

$$G = (1 + \frac{40k}{R_3})$$

$$R_1 = 4 k 7$$
$$R_2 = 100k$$
$$R_3 = 4k$$

Figure 10.14 Flowmeter signal conditioning circuit

Since it is the differential pressure across the orifice plate that is related to the flow rate, the transducer outputs were buffered with unity-gain operational amplifiers, OP-07CN, such as RS 301-892, as shown. The output voltage from each buffer amplifier was then input to a precision instrumentation amplifier,

INA101HP, such as RS 636-227. The final output voltage is in direct proportion to the pressure difference across the orifice plate. A calibration of the circuit gave a linear relationship between the volumetric flow rate and the square-root of the voltage output, i.e. $Q = 0.283\sqrt{V}$ litres/s, with the flow rate ranging between 0 and 0.35 litres/s.

(ii) Coolant Bulk Mean Temperature

The coolant bulk mean temperature was controlled by passing the coolant through a coil which was immersed in a controlled temperature, hot bath heat exchanger. The bath temperature was continuously monitored and heat was supplied using a standard electric kettle type of immersion heater. Choosing a dead band of $1.0°C$, a simple ON/OFF control strategy was adopted to stabilise the bath temperature at the desired set point. The power switching interface is shown in figure 10.15.

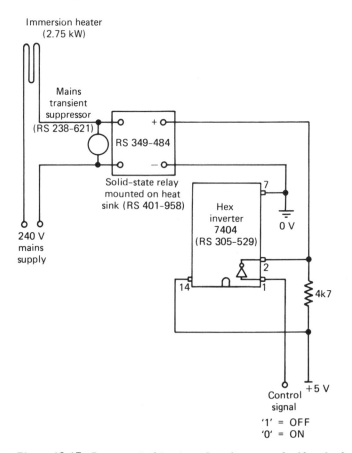

Figure 10.15 Power switching interface for control of hot bath temperature

(iii) Temperature Measurement

Standard K-type thermocouples were used to measure the temperatures at the locations, T_1 to T_3 in figure 10.13. The thermocouple signal conditioning interface employed the amplifier described in section 2.2, AD595. To reduce costs, however, the four thermocouple inputs were sampled in sequence using two multiplexers, DG508ACJ, such as RS 309-571. The temperature measurement interface is shown in figure 10.16.

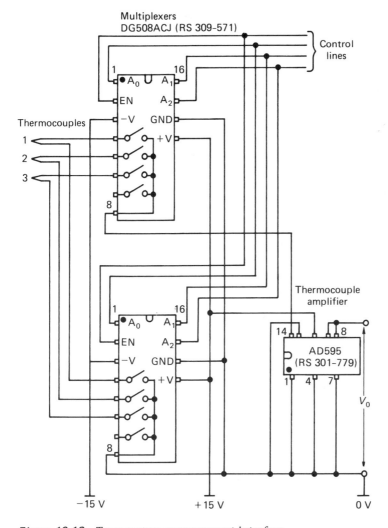

Figure 10.16 Temperature measurement interface

Note that two individual multiplexers are essential because none of the thermocouple output signal lines can be commoned. There would be a tendency, in fact, to corrupt the signal levels if any of the lines are commoned. Both multiplexers are operated from the same digital control lines from the microcomputer however. The four control lines include an output enable, EN, and three additional lines which can activate any one of eight switches on each multiplexer depending on the control signals on these lines.

(iv) The Microcomputer Interface

The microcomputer interface system chosen for this application was Control Universal's 'BEEBEX Cube', which is compatible with the BBC microcomputer. BEEBEX is a rack-mounted system which plugs into the 1 MHz expansion bus of the BBC micro. The system takes standard Eurocards including A/D converters, D/A converters, digital I/O and heavy-duty industrial opto-isolated I/O. The expansion card used with BEEBEX was the CUBAN-8 analogue and digital interface. This card features:

16 analogue input channels
1 analogue input channel
16 digital I/O lines in a VIA with 4 control lines
100 μs conversion time
8-bit accuracy, $\pm$ the LSB
input voltage range of 0 to 2.5 volts

The card has a link selectable base address which is normally set to &D800. Access to the card was provided by means of a 'Real Time BASIC' programming language. This was added to the microcomputer in the form of a 'sideways ROM', fitted in one of the spare sockets under the keyboard.

Four of the digital I/O lines were used to control the thermocouple multiplexers and a fifth line used to control the power switching to the hot bath heater. Only two of the analogue input channels were used, these being allocated to the flowmeter input signal and the input signal from the multiplexed thermocouples.

The data-acquisition and control program is given below:

```
10 REM Boiling heat transfer experiment
20 INPUT"Enter the hot bath set point temp, deg C ";SP
30 CLS
35 REM set the base address for the card
40 DBASE=&DB00
45 REM set channel 8, the relay control line, for output
50 OUTCH 8
55 REM set channels 0-4, the multiplexer control lines, for output
60 OUTCH 0 TO 4
65 REM switch multiplexer to select T1,T2 and T3 in sequence
70 FOR N%=0 TO 4 STEP 2
80 PORT 0=N%
85 REM time delay
```

```
 90 FOR C=1 TO 1500:NEXT C
100 SUM%=0
105 REM sample the thermocouple input channel
106 REM average of 100 values
110 FOR A=1 TO 100
120 SUM%=SUM%+ADC(65)
125 REM short delay
130 FOR B=1 TO 10:NEXT B
140 NEXT A
145 REM heater ON/OFF control statements
150 IF N%=0 AND SUM%/100<SP THEN CH8=1
160 IF N%=0 AND SUM%/100>SP+1 THEN CH8=0
170 PRINT TAB(8,N%);" T";N%/2+1;" :- ";SUM%/100;" deg C"
175 REM sample flowmeter signal input channel
176 REM average of 500 values
180 FOR D=1 TO 500
190 SUMFLO%=SUMFLO%+ADC(64)/256
200 NEXT D
205 REM Flowmeter Calibration Equation
210 FLO=INT(0.0282883*(SQR(SUMFLO%/500)-0.0046597)*10000+5)/10000
220 PRINT TAB(8,6);"Flow :- ";FLO;" Litres/sec"
230 SUMFLO%=0
240 GOTO 70
```

Real Time BASIC is initiated with the command *RTBASIC and is terminated with *UNREAL.

The language is an extension of the standard BBC BASIC and it provides additional commands which are particularly suited to control applications. Although these commands are not used in the example given, they have names which relate directly to their control functions, e.g. TURNON, TURNOFF, DELAY, etc. This makes the language easier to understand in a control context and enables the user to develop more efficient coding.

10.4 DIGITAL CONTROL OF A d.c. MOTOR

In order to control a process, measured information referring to the value of the process variables, the status of the process equipment and the economic performance of the system must be obtained. This data is then used to make control decisions according to some chosen strategy. Because of its speed of computation and decision making capability, the digital computer is a natural choice for plant monitoring and control.

In a dynamic environment such as the 'real world', disruptive forces are frequently encountered. The process variable must therefore be monitored and reset regularly in a feedback control loop, illustrated in figure 10.17 as a classical block diagram.

Direct digital control, abbreviated to DDC, has now virtually replaced the conventional analogue systems and the computer-based controller performs calculations to simulate the operation of the analogue control elements. Being a digital system, the process variable is sampled periodically at a sampling frequency which is sufficient to approximate the performance of that of the continuous analogue control.

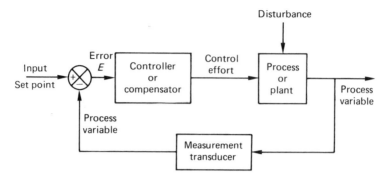

Figure 10.17 A single feedback control loop

The concept of direct digital control is effectively illustrated by the use of a microcomputer as an on-line real-time three term process controller applied to controlling the speed of a d.c. servo motor. Such an application using standard off-the-shelf RS components is invaluable in examining the response of a system for chosen controller settings.

Details of each constituent part of the control loop with the necessary interfacing follow.

1. Process

This is effectively a 12 V precision d.c. servo motor, RS 336-292, connected to a 5:1 ratio precision gearbox, RS 336-220, driving a high inertial load. The inertial

Figure 10.18 d.c. servo motor driving an inertial load

load consists of an aluminium drum, 40 mm diameter by 40 mm long, which is mounted on a steel spindle that runs in brass bushes. These are held between two support plates with the drive from the gearbox made via a universal joint. The configuration is shown in figure 10.18 and the drum provides a means of supplying stepped loads in order to observe the controller and the motor response.

2. Measurement Transducer

A magnetic sensor as illustrated in figure 2.2 is used in conjunction with the tachometer circuit presented in figure 2.3 to measure the drum rotational speed. The signal pulses are generated from four metal blocks attached to the side of the aluminium drum. The resulting output voltage from the tachometer is duly amplified using an op-amp non-inverting circuit as shown in figure 2.30b. This gives a voltage of 10 V to correspond to a drum speed of 2000 rev/min. This value is then suitable for inputting directly to the chosen analogue-to-digital converter.

3. Controller

Use is made of an AMSTRAD 1512PC with an interface card which has the following specification:
(a) A 16 channel multiplexed sample and hold 12-bit resolution A/D converter with an overall conversion time of 35 μs and an input voltage range of 0 V to +10 V.
(b) Three 8-bit programmable input/output ports A, B and C.
The output speed is read on channel 1 of the multiplexer and the control effort value (0-255), generated by the control strategy algorithm, is output to port B.
The I/O address space used by the card to utilise two 8255 PPIs (see section 5.3) is as follows:
The base address (BA) is set as 700 hex.
PPI1 — for use with the ADC:

Address	Function
BA + 00	Input 8 LSBs from ADC
BA + 01	Input 4 MSBs (lower nibble) from ADC
BA + 02	Lower nibble — start ADC conversion by toggling bit 0 low/high while keeping bit 1 high
	Upper nibble — select multiplexer channel
BA + 03	Control word for PPI1 (= 92 hex)

PPI2 — for providing the three digital I/O ports:

Address	*Function*
BA + 08	Port A
BA + 09	Port B — set to output for control effort value
BA + 10	Port C
BA + 11	Control word for PPI2 (= 90 hex)

Hence the ADC is enabled for channel 1 with:

    OUT &H702,&H12
    OUT &H702,&H13

There is no 'end of conversion' sensing facility. It is probably unnecessary when programming in high level BASIC but it is better to include a small delay to ensure that the conversion is complete. The 12-bit value representing the output speed is obtained by reading addresses &H700 and &H701, that is

    $I = 256*(INP(\&H701) \text{ AND } 15) + INP(\&H700)$

from which the corresponding speed in rev/min is

    $PV = 2000*(I/4095)$

This gives an error at any particular time of:

    $E = SP - PV$

4. The Control Effort Interface

The 8-bit number generated by the chosen digital control strategy to represent the control effort has to be converted into a voltage within the range 0 to 12 V. This should also have a suitable power driving capability to suit the motor. The d.c. servo motor control module, such as RS 591-663, is suitable for this purpose and requires an input voltage of 0–4 V to vary the motor speed over its full range. The voltage is generated by taking the 8-bit output from port B of PPI2 into a ZN428E D/A converter, such as RS 303-523, and amplifying the output to give a maximum of 4 V. The corresponding wiring diagram is shown in figure 10.19.

5. The Digital Control Algorithm

The normal control technique adopted in industrial process control applications is the so-called three term or PID control strategy which is based on the error value E. It is formulated as

$$\text{Control Effort} = KE + (K/T_i) \int E \, dt + KT_d \, dE/dt$$

where K, T_i and T_d are the controller settings for gain, integral time or reset and derivative time.

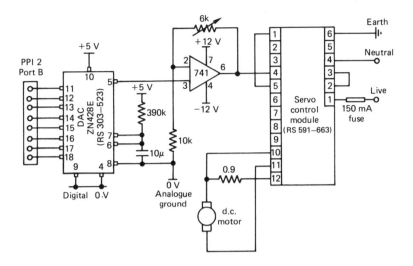

Figure 10.19 d.c. servo motor power driving circuit

It is easily translated into a digital algorithm based on a chosen interval sampling time of DT. The integral term is then obtained from the summation of the error value and the derivative term from the finite change in the error over the time interval DT.

The choice of step length is dependent upon the actual application but an empirical rule is to use at least one-tenth of the closed loop control settling time. For the application considered, the motor with inertial load takes about 2 seconds to achieve a steady output speed of around 1200 rev/min with a step input from rest. A suitable time increment of 0.2 second or less would therefore be appropriate. Since the TIMER function in BASICA is specified as having a resolution of about one-tenth of a second, DT is taken as 0.1 second in the digital control algorithm. The program below in BASICA inputs required set point speed and controller settings and controls the motor speed over a specified time period.

```
10  REM set up both 8255 PPI ports as necessary
20  OUT &H703,&H92
30  OUT &H70B,&H90
40  REM switch motor off
50  OUT &H709,&H0
60  CLS
70  LOCATE 3,10
80  PRINT "This program illustrates the characteristics of a servo motor"
90  LOCATE 5,10
100 PRINT "in response to a three term PID control action"
110 LOCATE 11,10
120 INPUT "Required speed in rev/min = ";SP
130 LOCATE 13,10
140 INPUT "Controller gain = ";K
150 LOCATE 15,10
```

```
160 INPUT "Controller integral time in seconds = ";TI
170 LOCATE 17,10
180 INPUT "Controller derivative time in seconds = ";TD
190 LOCATE 19,10
200 INPUT "Duration of required run time in seconds = ";D
210  REM *****************************************************
220 DT=.1:SUM=0:EP=0
230 IC=K*DT/TI:DC=K*TD/DT
240  REM zero timer for run duration
250 T1=TIMER
260  REM start of control loop sequence
270 WHILE (TIMER-T1)<D
280 T2=TIMER
290 OUT &H702,&H12
300 OUT &H702,&H13
310 FOR J=1 TO 5:NEXT J
320 A=INP(&H701):B=INP(&H700)
330 I=256*(A AND 15)+B
340 PV=2000*(I/4095)
350 E=SP-PV
360 SUM=SUM+E
370 VOUT%=(K*E)+(IC*SUM)+(DC*(E-EP))
380 IF VOUT%<0 THEN VOUT%=0
390 IF VOUT%>255 THEN VOUT%=255
400  REM output control effort value to DAC
410 OUT &H709,VOUT%
420 EP=E
430  REM check for control loop contained within DT seconds
440 IF (TIMER-T2)<.1 THEN GOTO 440
450 WEND
460 OUT &H709,0
470 END
```

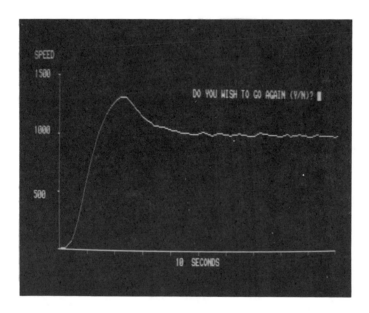

Figure 10.20   *Graphics output for d.c. servo motor speed response*

Typical controller setting values to achieve a reasonable response are:

$K = 0.3$, $T_i = 0.6$ second, $T_d = 0.15$ second

Adding a graphical output to the program greatly enhances its value in demonstrating the effect of various settings on the system response and a screen display from the AMSTRAD 1512 when running with the above values to achieve a set point of 1000 rev/min is shown in figure 10.20.

The example given might be thought of as 'illustrative', or as a 'demonstration'. The principles employed are fundamental nonetheless and are the same as those which would be applied in an industrial context to implement PID control of virtually any item of plant or a complete process.

REFERENCES

Control Universal Ltd, *Cube and the BBC Microcomputer, Real Time BASIC, BBC Version*, Control Universal, Cambridge, 1986.

Eastop, T. D. and McConkey, A., *Applied Thermodynamics for Engineering Technologists*, 4th edn, Longman, 1986.

Jervis, J. M., A computer based engine indicator system, *Conf. on Developments in Measurements and Instrumentation in Engineering*, Mechanical Engineering Publications, 1985.

Rodgers, G. F. C. and Mayhew, Y. R., *Engineering Thermodynamics Work and Heat Transfer*, 3rd edn, Longman, 1980.

Appendix I: Polynomial Curve Fitting—Method of Least Squares

The least squares method of curve fitting involves an approximating function, which is chosen such that the function will pass as close as possible, but not necessarily through, all of the original data points. The numerical technique involves processing to ensure that the sum of the squares of the differences between the approximating function and the actual data points is minimised. The problem then resolves to one of choosing an appropriate form for the approximating function to best fit the data.

Let the approximating function be $g(x)$.

The difference between the approximating function, $g(x_i)$ and the 'true' curve, at the point (x_i, y_i) is:

$$g(x_i) - y_i \tag{i}$$

This difference is then evaluated over a number of discrete points, the square of the differences calculated and then summed for all of the points:

$$M = \sum_{i=1}^{i=n} \{g(x_i) - y_i\}^2 \tag{ii}$$

where n is the number of points considered.

The approximating function may be cast as a linear combination of suitable sub-functions, that is:

$$g(x) = C_0 + C_1 g_1(x) + \ldots + C_k g_k(x) \tag{iii}$$

where $C_0, C_1, \ldots, C_k$ are as yet unknown constants and k is the selected order of the function.

The condition for a minimum M is:

$$\frac{\partial M}{\partial C_0} = \frac{\partial M}{\partial C_1} = \ldots \frac{\partial M}{\partial C_k} = 0$$

Substituting (iii) into (ii) gives:

$$M = \sum_{i=1}^{i=n} \{C_0 + C_1 g_1(x_i) + \ldots + C_k g_k(x_i) - y_i\}^2 \tag{iv}$$

$$\frac{\partial M}{\partial C_0} = 2 \sum_{i=1}^{i=n} \{C_0 + C_1 g_1(x_i) + \ldots + C_k g_k(x_i) - y_i\} = 0 \tag{v}$$

Similarly:

$$\frac{\partial M}{\partial C_1} = 2 \sum_{i=1}^{i=n} \{C_0 + C_1 g_1(x_i) + \ldots + C_k g_k(x_i) - y_i\} g_1(x_i) = 0 \tag{vi}$$

and

$$\frac{\partial M}{C_k} = 2 \sum_{i=1}^{i=n} \{C_0 + C_1 g_1(x_i) + \ldots + C_k g_k(x_i) - y_i\} g_k(x_i) = 0 \tag{vii}$$

Equations (v), (vi) and (vii) may be re-written as:

$$\sum_{i=1}^{i=n} C_0 + C_1 \sum_{i=1}^{i=n} g_1(x_i) + \ldots + C_k \sum_{i=1}^{i=n} g_k(x_i) = \sum_{i=1}^{i=n} y_i \tag{viii}$$

$$C_0 \sum_{i=1}^{i=n} g_1(x_i) + C_1 \sum_{i=1}^{i=n} g_1^2(x_i) + \ldots + C_k \sum_{i=1}^{i=n} g_k(x_i) g_1(x_i) = \sum_{i=1}^{i=n} y_i g_1(x_i) \tag{ix}$$

$$C_0 \sum_{i=1}^{i=n} g_k(x_i) + C_1 \sum_{i=1}^{i=n} g_1(x_i) g_k(x_i) + \ldots + C_k \sum_{i=1}^{i=n} g_k^2(x_i) = \sum_{i=1}^{i=n} y_i g_k(x_i) \tag{x}$$

The summations can be obtained from the experimental data and this gives a set of simultaneous equations in $C_0, C_1, \ldots, C_k$, which can be solved using any suitable numerical method.

Example 1

A straight line fit is of the general form:

$$y = C_0 + C_1 x = g(x)$$

This is one of the simplest cases, with $g_1 = x$ and $g_2 = g_3 = \ldots = g_k = 0$.
From equations (viii) and (ix):

$$n C_0 + C_1 \sum_1^n x = \sum_1^n y$$

$$C_0 \sum_1^n x + C_1 \sum_1^n x^2 = \sum_1^n y x$$

In matrix form:

$$\begin{bmatrix} n & \sum_1^n x \\ \sum_1^n x & \sum_1^n x^2 \end{bmatrix} \times \begin{bmatrix} C_0 \\ C_1 \end{bmatrix} = \begin{bmatrix} \sum_1^n y \\ \sum_1^n yx \end{bmatrix}$$

Example 2

A cubic is of the general form:

$$y = C_0 + C_1 x + C_2 x^2 + C_3 x^3 = g(x)$$

In this case, $g_1 = x$, $g_2 = x^2$, $g_3 = x^3$ and $g_4 = g_5 = \ldots = g_k = 0$.
From equations (viii), (ix) and (x):

$$n C_0 + C_1 \sum_1^n x + C_2 \sum_1^n x^2 + C_3 \sum_1^n x^3 = \sum_1^n y$$

$$C_0 \sum_1^n x + C_1 \sum_1^n x^2 + C_2 \sum_1^n x^2 x + C_3 \sum_1^n x^3 x = \sum_1^n y x$$

$$C_0 \sum_1^n x^2 + C_1 \sum_1^n x x^2 + C_2 \sum_1^n x^2 x^2 + C_3 \sum_1^n x^3 x^2 = \sum_1^n y x^2$$

$$C_0 \sum_1^n x^3 + C_1 \sum_1^n x x^3 + C_2 \sum_1^n x^2 x^3 + C_3 \sum_1^n x^3 x^3 = \sum_1^n y x^3$$

Or in matrix form:

$$\begin{bmatrix} n & \sum_1^n x & \sum_1^n x^2 & \sum_1^n x^3 \\ \sum_1^n x & \sum_1^n x^2 & \sum_1^n x^3 & \sum_1^n x^4 \\ \sum_1^n x^2 & \sum_1^n x^3 & \sum_1^n x^4 & \sum_1^n x^5 \\ \sum_1^n x^3 & \sum_1^n x^4 & \sum_1^n x^5 & \sum_1^n x^6 \end{bmatrix} \times \begin{bmatrix} C_0 \\ C_1 \\ C_2 \\ C_3 \end{bmatrix} = \begin{bmatrix} \sum_1^n y \\ \sum_1^n y x \\ \sum_1^n y x^2 \\ \sum_1^n y x^3 \end{bmatrix}$$

With the cubic approximating function, it can be seen that an ordered pattern emerges in the simultaneous equations to be solved. The same pattern continues for higher order polynomials and advantage can be taken of this to write a generalised program to solve for polynomials of up to any prescribed order. The only proviso is that the number of experimental data points to be included must be at least one more than the order of polynomial specified.

In the following program, the matrix inversion is carried out using the Gauss elimination technique and a graphics routine is included as a check on the 'goodness of fit'. Any polynomial of up to order 10 may be specified as the approximating function.

```
10 REM
20 REM This program fits a polynomial to N sets of experimental data
30 REM The order of the polynomial can be up to order 10 if required
40 REM
50 CLS:SCREEN 2
60 DIM A(10,20),AINV(10,10)
70 DIM C(10),B(10),X(40),Y(40),XN(40),YN(40)
80 REM
90 PRINT:PRINT:PRINT"              POLYNOMIAL FITTING PROGRAM"
100 PRINT:PRINT:PRINT" Enter the degree of the polynomial to be fitted"
110 PRINT:PRINT"        ( UP TO ORDER 10 ONLY )                 ";
120 INPUT M
130 PRINT:PRINT:PRINT" Enter the number of data points ";
140 INPUT N
150 IF (N<M OR N=M) GOTO 130
160 PRINT:PRINT:PRINT" Enter the data points in pairs,"
170 PRINT" X-coordinate first, Y-coordinate second"
180 PRINT:PRINT" Separate the data with a comma "
190 PRINT:PRINT
200 FOR K=1 TO N:INPUT X(K),Y(K):NEXT K
210 PRINT:PRINT" Data entered, Computation in progress "
220 REM
230 REM initialisation of matrices
240 REM
250 MAX=2*M
260 REM
270 FOR I=1 TO M
280 FOR J=1 TO MAX
290 A(I,J)=0
300 NEXT J:NEXT I
310 A(1,1)=N
320 FOR I=2 TO M
330 FOR K=1 TO N
340 A(I,1)=A(I,1)+X(K)^(I-1)
350 NEXT K:NEXT I
360 FOR J=2 TO M
370 FOR K=1 TO N
380 A(M,J)=A(M,J)+X(K)^(M+J-2)
390 NEXT K:NEXT J
400 FOR J=2 TO M
410 FOR I=1 TO (M-1)
420 A(I,J)=A((I+1),(J-1))
430 NEXT I:NEXT J
440 FOR I=1 TO M
450 A(I,(M+I))=1
460 B(I)=0:C(I)=0:NEXT I
470 FOR I=1 TO M
480 FOR K=1 TO N
490 B(I)=B(I)+Y(K)*X(K)^(I-1)
500 NEXT K:NEXT I
510 REM
520 REM Matrices now initialised
530 REM
540 L=1
550 FOR J=1 TO (M-1)
560 L=L+1
```

```
570 FOR I=L TO M
580 CONST=-(A(I,J)/A(J,J))
590 FOR K=J TO MAX
600 A(I,K)=A(I,K)+CONST*A(J,K)
610 NEXT K:NEXT I:NEXT J
620 L=0
630 FOR J=M TO 2 STEP -1
640 L=L+1
650 FOR I=(M-L) TO 1 STEP -1
660 CONST=-(A(I,J)/A(J,J))
670 FOR K=MAX TO 2 STEP -1
680 A(I,K)=A(I,K)+CONST*A(J,K)
690 NEXT K:NEXT I:NEXT J
700 FOR I=1 TO M
710 ANOR=A(I,I)
720 FOR J=1 TO MAX
730 A(I,J)=A(I,J)/ANOR
740 NEXT J:NEXT I
750 K=0
760 FOR J=(M+1) TO MAX
770 K=K+1
780 FOR I=1 TO M
790 AINV(I,K)=A(I,J)
800 NEXT I:NEXT J
810 REM
820 REM Matrix inversion complete
830 REM
840 FOR I=1 TO M
850 FOR J=1 TO M
860 C(I)=C(I)+AINV(I,J)*B(J)
870 NEXT J:NEXT I
880 CLS
890 PRINT:PRINT:PRINT" The Polynomial Approximation is :- "
900 PRINT:PRINT" Co + C1*X + C2*x^2 + - - - - - - + C9*X^9"
910 PRINT:PRINT" The polynomial constants are :- "
920 PRINT:PRINT
930 FOR I=1 TO M
940 PRINT:PRINT"    C";I-1;" = ";C(I)
950 NEXT I
970 XMAX=X(1):YMAX=Y(1)
980 FOR K=2 TO N
990 IF X(K)>XMAX THEN XMAX=X(K)
1000 IF Y(K)>YMAX THEN YMAX=Y(K)
1010 NEXT K
1020 FOR K=1 TO N
1030 XN(K)=X(K)/XMAX
1040 YN(K)=Y(K)/YMAX
1050 NEXT K
1060 PRINT:PRINT" Press enter to continue ";
1070 INPUT C$
1080 REM
1090 REM Graphics output
1100 REM
1101 CLS
1110 LINE (100,160)-(500,160)
1120 LINE (100,160)-(100,20)
1130 FOR K=1 TO N
1140 PSET (100+XN(K)*400,160-YN(K)*140)
1150 NEXT K
1160 LOCATE 2,15:PRINT" Normalised input data "
1170 LOCATE 22,12:PRINT" press enter to continue ";
1180 INPUT C$
1190 LOCATE 22,12:PRINT"                              "
1200 LOCATE 2,15:PRINT" Polynomial Aproximation"
1210 C0=C(1):C1=C(2):C2=C(3):C3=C(4):C4=C(5):C5=C(6)
1220 C6=C(7):C7=C(8):C8=C(9):C9=C(10)
1230 FOR K=0 TO 1 STEP .002
```

```
1240 YP=(C0+C1*K*XMAX+C2*K*K*XMAX*XMAX+C3*K^3*XMAX^3+C4*K^4*XMAX^4+C5*K^5*XMAX^
5+C6*K^6*XMAX^6+C7*K^7*XMAX^7+C8*K^8*XMAX^8+C9*K^9*XMAX^9)/YMAX
1250 PSET (100+K*400,160-YP*140)
1260 NEXT K
1270 LOCATE 22,2
1280 END
```

The program given in BASICA may be run on an IBM-PC. To convert the coding to BBC BASIC line 50 should be altered to:

50 CLS:MODE 7

In addition, lines 1100 to the end of the program should be replaced with:

```
1100 REM
1110 MODE4
1120 MOVE 50,50
1130 DRAW 50,850
1140 MOVE 50,50
1150 DRAW 1250,50
1160 MOVE 50,50
1170 FOR K=1 TO N
1180 PLOT 69,(XN(K)*1200+50),(YN(K)*800+50)
1190 NEXT
1200 PRINTTAB(5,2)"PRESS RETURN TO CONTINUE"
1210 INPUT QS
1220 C0=C(1):C1=C(2):C2=C(3):C3=C(4):C4=C(5):C5=C(6):C6=C(7)
1230 C7=C(8):C8=C(9):C9=C(10)
1240 MOVE 50,50
1250 FOR K=0 TO 1.05 STEP 0.05
1260 YP=(C0+C1*K*XMAX+C2*K*K*XMAX*XMAX+C3*K↑3*XMAX↑3+C4*K↑4*XMAX↑4+C5*K↑
5*XMAX↑5+C6*K↑6*XMAX↑6+C7*K↑7*XMAX↑7+C8*K↑8*XMAX↑8+C9*K↑9*XMAX↑9)/YMAX
1270 DRAW K*1200+50,YP*800+50
1280 NEXT
1290
1300 END
```

Appendix II:
Instruction Set for the 6502
Microprocessor

SINGLE COMMANDS

CLC   Clear carry flag
SEC   Set carry flag
CLD   Clear decimal flag
SED   Set decimal flag
NOP   No operation
BRK   Break

DATA TRANSFER WITH ACCUMULATOR

LDA   Load accumulator – immediate mode, zero page, absolute and indexed
STA   Store accumulator – zero page, absolute and indexed modes

INDEXED REGISTERS

LDX   Load X register – immediate mode, zero page and absolute
LDY   Load Y register – immediate mode, zero page and absolute
STX   Store X – zero page and absolute modes
STY   Store Y – zero page and absolute modes
TAX   Transfer accumulator to X register
TXA   Transfer X register to accumulator
TAY   Transfer accumulator to Y register
TYA   Transfer Y register to accumulator

ARITHMETIC AND LOGIC COMMANDS

ADC   Add with carry to accumulator – immediate mode, zero page, absolute and indexed

SBC   Subtract with carry from accumulator – immediate mode, zero page, absolute and indexed

AND   AND with accumulator – immediate mode, zero page, absolute and indexed

ORA   OR with accumulator – immediate mode, zero page, absolute and indexed

EOR   Exclusive OR with accumulator – immediate mode, zero page and indexed

INCREMENT OR DECREMENT MEMORY, X OR Y

INC   Increment memory by one – zero page, absolute and indexed

DEC   Decrement memory by one – zero page, absolute and indexed

INX   Increment X register

DEX   Decrement X register

INY   Increment Y register

DEY   Decrement Y register

COMPARE COMMANDS

CMP   Compare memory with accumulator – immediate mode, zero page, absolute and indexed

CPX   Compare memory with X – immediate mode, zero page, absolute and indexed

CPY   Compare memory with Y – immediate mode, zero page, absolute and indexed

BRANCH INSTRUCTIONS

BCC   Branch on carry flag clear (C = 0)

BCS   Branch on carry flag set (C = 1)

BEQ   Branch on zero flag set (Z = 1)

BNE   Branch on zero flag clear (Z = 0)

BMI   Branch on negative flag set (N = 1)

BPL   Branch on negative flag clear (N = 0)

JUMP INSTRUCTIONS

JMP   Jump to absolute address
JSR   Jump to absolute address of subroutine
RTS   Return from subroutine

OTHER COMMANDS

ASL, LSR, ROR, ROL	bit movement through memory or accumulator
PHA, PLA, PHP, PLP	using the stack
CLI, SEI, RTI	using the interrupts

ADDRESSING MODES

Mnemonics	Immediate	Zero page	Absolute	Implied	Relative
ADC	69	65	6D		
AND	29	25	2D		
ASL		06	0E		
BCC					90
BCS					B0
BEQ					F0
BIT		2C			
BMI					30
BNE					D0
BPL					10
BRK				00	
BVC					50
BVS					70
CLC				18	
CLD				D8	
CLI				58	
CLV				B8	
CMP	C9	C5	CD		
CPX	E0	E4	EC		
CPY	C0	C4	CC		
DEC		C6	CE		
DEX				CA	
DEY				88	
EOR	49	45	4D		
INC		E6	EE		
INX				E8	

Mnemonics	Immediate	Zero page	Absolute	Implied	Relative
INY				C8	
JMP			4C		
JSR			20		
LDA	A9	A5	AD		
LDX	A2	A6	AE		
LDY	A0	A4	AC		
LSR		46	4E		
NOP				EA	
ORA	09	05	0D		
PHA				48	
PHP				08	
PLA				68	
PLP				28	
ROL		26	2E		
ROR		66	6E		
RTI				40	
RTS				60	
SBC	E9	E5	ED		
SEC				38	
SED				F8	
SEI				78	
STA		85	8D		
STX		86	8E		
STY		84	8C		
TAX				AA	
TAY				A8	
TYA				98	
TSX				BA	
TXA				8A	
TXS				9A	

Appendix III:
Glossary of Terms

a.c.	alternating current
bandwidth	frequency range over which gain is constant
baud	serial transmission rate
bit	binary digit
buffer	high input impedance, low output impedance amplifier
byte	eight bits
db	decibels
d.c.	direct current
e.m.f.	electro-motive force
gain	ratio of output to input
hex	hexadecimal
nibble	four bits, i.e. half a byte
o/p	output
port	8-bit parallel input/output channel
r.m.s.	root mean square
ACR	Auxiliary Control Register
A/D	Analogue to Digital
ADC	Analogue-to-Digital Converter
ADFS	Advanced Disc Filing System
ALU	Arithmetic and Logic Unit
ASCII	American Standard Code for Information Interchange
BBC	British Broadcasting Corporation
BDC	Bottom Dead Centre
BS	British Standard
CAD	Computer Aided Design
CGA	Colour Graphics Adaptor
CMOS	Complimentary Metal Oxide Semi-conductor

CONVERT	start conversion
CPU	Central Processor Unit
CS	Channel Select
CTA	Constant Temperature Anemometer
D/A	Digital to Analogue
DAC	Digital-to-Analogue Converter
DDR	Data Direction Register
DFS	Disc Filing System
DIL	Dual In Line
DIN	Deutsche Industrie Norm (European standard for electrical connectors)
DMA	Direct Memory Access
DOS	Disc Operating System
DR	Data Register
DVM	Digital Volt Meter
ECD	Enhanced Colour Display
EEPROM	Electrically Erasable Programmable Read Only Memory
EGA	Enhanced Graphics Adaptor
EIA	American Electrical Industries Association
EOC	End Of Conversion
EPROM	Erasable Programmable Read Only Memory
GBW	product of Gain and Bandwidth
IBM	International Business Machines
IC	Integrated Circuit
IEEE	Institute of Electrical and Electronics Engineers
IEEE-488	parallel communication standard
IER	Interrupt Enable Register
IFR	Interrupt Flags Register
IMEP	Indicated Mean Effective Pressure
I/O	Input/Output
I/P	Current to Pressure
IRQ	Interrupt ReQuest
LED	Light Emitting Diode
LSB	Least Significant Bit
LVDT	Linearly Variable Differential Transformer
MDPA	Monochrome Display Printer Adaptor
MSB	Most Significant Bit
MS-DOS	MicroSoft – Disc Operating System
MUX	Multiplexer
PB	Proportional Band
PC	Personal Computer, Program Counter
PC-DOS	Personal Computer – Disc Operating System
PCR	Peripheral Control Register
PID	Proportional, Integral, Derivative

PPI	Programmable Peripheral Interface
PS	Personal System
PV	Process Variable
RAM	Random Access Memory
RC	Resistive, Capacitive
ROM	Read Only Memory
RS	Radio Spares
RS232	
RS232C	serial communication standard
RS422 and RS423	
SC	Start Conversion
SCD	Standard Colour Display
SCR	Silicon Controlled Rectifier
S/H	Sample and Hold
SI	Système International (metric system of units)
SP	Set Point, Stack Pointer
SRQ	Service ReQuest (management control line)
STATUS	end of conversion
TDC	Top Dead Centre
TTL	Transistor Transistor Logic
VGA	Video Graphics Adaptor
V/I	Voltage to Current
VIA	Versatile Interface Adaptor
VLSI	Very Large Scale Integration

Appendix IV:
Hexadecimal Conversion
Table

Hex	0	1	2	3	4	5	6	7	8	9	A	B	C	D	E	F
0	0	1	2	3	4	5	6	7	8	9	10	11	12	13	14	15
1	16	17	18	19	20	21	22	23	24	25	26	27	28	29	30	31
2	32	33	34	35	36	37	38	39	40	41	42	43	44	45	46	47
3	48	49	50	51	52	53	54	55	56	57	58	59	60	61	62	63
4	64	65	66	67	68	69	70	71	72	73	74	75	76	77	78	79
5	80	81	82	83	84	85	86	87	88	89	90	91	92	93	94	95
6	96	97	98	99	100	101	102	103	104	105	106	107	108	109	110	111
7	112	113	114	115	116	117	118	119	120	121	122	123	124	125	126	127
8	128	129	130	131	132	133	134	135	136	137	138	139	140	141	142	143
9	144	145	146	147	148	149	150	151	152	153	154	155	156	157	158	159
A	160	161	162	163	164	165	166	167	168	169	170	171	172	173	174	175
B	176	177	178	179	180	181	182	183	184	185	186	187	188	189	190	191
C	192	193	194	195	196	197	198	199	200	201	202	203	204	205	206	207
D	208	209	210	211	212	213	214	215	216	217	218	219	220	221	222	223
E	224	225	226	227	228	229	230	231	232	233	234	235	236	237	238	239
F	240	241	242	243	244	245	246	247	248	249	250	251	252	253	254	255

Index